ELECTRICITY AND MAGNETISM.

By FLEEMING JENKIN, F.R.SS. L. & E. Professor of Engineering in the University of Edinburgh. New Edition, revised. Price 3*s.* 6*d.*

WORKSHOP APPLIANCES,

including Descriptions of the Gauging and Measuring Instruments, the Hand Cutting-Tools, Lathes, Drilling, Planing, and other Machine Tools used by Engineers. By C. P. B. SHELLEY, Civil Engineer, Hon. Fellow and Professor of Manufacturing Art and Machinery at King's College, London. With 209 Figures engraved on Wood. Price 3*s.* 6*d.*

PRINCIPLES OF MECHANICS.

By T. M. GOODEVE, M.A. Barrister-at-Law, Lecturer on Applied Mechanics at the Royal School of Mines. With 208 Figures and Diagrams on Wood. Price 3*s.* 6*d.*

TEXT-BOOKS, NOW PUBLISHED, EDITED BY C. W. MERRIFIELD, F.R.S.—

QUANTITATIVE CHEMICAL ANALYSIS.

By T. E. THORPE, F.R.S.E. Ph.D. Professor of Chemistry in the Andersonian University, Glasgow. With 88 Figures on Wood. Price 4*s.* 6*d.*

INTRODUCTION TO THE STUDY OF ORGANIC CHEMISTRY;

The CHEMISTRY of CARBON and its COMPOUNDS. By HENRY E. ARMSTRONG. Ph.D. F.C.S. Professor of Chemistry in the London Institution. With 8 Figures on Wood. Price 3*s.* 6*d.*

QUALITATIVE CHEMICAL ANALYSIS AND LABORATORY PRACTICE.

By T. E. THORPE, Ph.D. F.R.S.E. Professor of Chemistry in the Andersonian University, Glasgow; and M. M. PATTISON MUIR, F.R.S.E. With Plate and 57 Figures on Wood. Price 3*s.* 6*d.*

TELEGRAPHY.

By W. H. PREECE, C.E. Divisional Engineer, Post-Office Telegraphs; and J. SIVEWRIGHT, M.A. Superintendent (Engineering Department) Post-Office Telegraphs. Price 3*s.* 6*d.*

RAILWAY APPLIANCES.

Including Permanent Way, Points and Crossings, Stations and Station Arrangements, Signals, Carriage and Waggon Stock, Breaks, and other Details of Railways.
By J. W. BARRY, Member of the Institution of Civil Engineers, &c. Price 3*s.* 6*d.*

TEXT-BOOKS PREPARING FOR PUBLICATION, TO BE EDITED BY
T. M. GOODEVE, M.A.

ECONOMICAL APPLICATIONS OF HEAT,

Including Combustion, Evaporation, Furnaces, Flues, and Boilers.
By C. P. B. SHELLEY, Civil Engineer, and Professor of Manufacturing Art and Machinery at King's College, London.
With a Chapter on the Probable Future Development of the Science of Heat, by C. WILLIAM SIEMENS, F.R.S.

THE STEAM ENGINE.

By T. M. GOODEVE, M.A. Barrister-at-Law, Lecturer on Mechanics at the Royal School of Mines.

SOUND AND LIGHT.

By G. G. STOKES, M.A. D.C.L. Fellow of Pembroke College, Cambridge; Lucasian Professor of Mathematics in the University of Cambridge; and Secretary to the Royal Society.

TEXT-BOOKS PREPARING FOR PUBLICATION, TO BE EDITED BY
C. W. MERRIFIELD, F.R.S.

PRACTICAL AND DESCRIPTIVE GEOMETRY, AND PRINCIPLES OF MECHANICAL DRAWING.

By C. W. MERRIFIELD, F.R.S. an Examiner in the Department of Public Education, and late Principal of the Royal School of Naval Architecture and Marine Engineering, South Kensington.

ELEMENTS OF MACHINE DESIGN.

With Rules and Tables for Designing and Drawing the Details of Machinery. Adapted to the use of Mechanical Draughtsmen and Teachers of Machine Drawing.
By W. CAWTHORNE UNWIN, B.Sc. Assoc. Inst. C.E. Professor of Hydraulic and Mechanical Engineering at Cooper's Hill College.

PHYSICAL GEOGRAPHY.

By the Rev. GEORGE BUTLER, M.A. Principal of Liverpool College; Editor of 'The Public Schools Atlas of Modern Geography.'

*** *To be followed by other works on other branches of Science.*

London, LONGMANS & CO.

TEXT-BOOKS OF SCIENCE

ADAPTED FOR THE USE OF

ARTISANS AND STUDENTS IN PUBLIC AND SCIENCE SCHOOLS

TELEGRAPHY

TELEGRAPHY

BY

W. H. PREECE, C.E.
Divisional Engineer, Post Office Telegraphs

AND

J. SIVEWRIGHT, M.A.
Superintendent (Engineering Department) Post Office Telegraphs

D. APPLETON AND CO.
NEW YORK
1876

PREFACE.

THIS text-book, although adapted for the use of students generally, is written specially for those numerous operators and artisans who are employed in the actual transmission of telegrams, and in the maintenance of telegraphs in England. Care has been taken to render it as far as possible independent of theory, and of little more than an elementary knowledge of Mathematics. The book is intended to serve as an introduction to the study of more advanced works upon the art and science of Telegraphy. Its dimensions have necessarily confined the authors almost entirely to the consideration of English Telegraphy, and compelled them to abandon the submarine cable branch of the subject. The systems described are those which have borne the test of continued experience and are more or less in practical use at the present day. Hence it is that the problems of quadruplex, multiplex, and other novel systems of Telegraphy have been omitted. And as the class for whom the book is specially written are not as a rule engaged in the application of the laws of currents to testing and experimental purposes, the discussion of Ohm's laws, and the apparatus depending upon them, are not dealt with. In fact, Professor Fleeming Jenkin's work on Electricity, published in this series, fills up the theoretical omissions in the book, and Mr. Culley's Handbook of Practical Telegraphy, to which, more than any other, this is intended to be an introduction, supplies all the practical omissions.

Those who take up this text-book with the idea that from it they are going to learn Telegraphy without any previous knowledge of Electricity, and without the opportunity of handling telegraph instruments, will probably be disappointed ; while those who have already acquired an elementary knowledge of Electricity, or are employed in Telegraphy, will, it is believed, find the work to supply a gap which the authors have often felt to exist.

Wherever the diagrams are drawn to scale they are so indicated by the proportion of the real size being given. Wherever this is not so the diagrams are either symbolical or simply illustrative without strict adherence to dimensions.

CONTENTS.

CHAPTER I.

ELECTRICAL TERMS.

CHAPTER II.

THE BATTERY.

CHAPTER III.

SIGNALLING INSTRUMENTS.

CHAPTER IV.

CIRCUITS.

CHAPTER V.

SPECIAL TELEGRAPHY.

CHAPTER VI.

CONSTRUCTION.

CHAPTER VII.

CONSTRUCTION (*continued*).

CHAPTER VIII.

FAULTS.

CHAPTER IX.

TESTING.

CHAPTER X.

COMMERCIAL TELEGRAPHY.

TELEGRAPHY.

CHAPTER I.

ELECTRICAL TERMS.

1. It is not intended that this book shall be a treatise on electricity. It is to be a Text-book of Telegraphy, dealing with the application of electricity to the conveyance of information to distant points beyond the reach of the ear and the eye. Electricity therefore being its main theme, a certain acquaintance with the elementary principles of that science must be assumed on the part of the reader, though every effort will be made to render the explanations given, as much as possible independent of such knowledge. The differences which exist among electricians with respect to the signification of many of the technical terms employed in connection with telegraphy render it necessary that the student at the outset should have a clear comprehension of the meaning of those which will be used in this work. Whether electricity be a fluid or a force, whether it be a form of matter or a form of energy, will not be discussed; the terms used will be rendered independent of theory.

2. Electricity is an agent pervading terrestrial and solar space, and is as universal in its effects as are heat and light. We are cognizant of its existence when we hear the roar of thunder and see the flash of lightning, but we do not know its particular form any more than we know that of heat or that of light. The sound of the thunder and the flash of the

lightning affect the ear and the eye—we hear the sound and see the light—but we do not assume the existence either of sound or of light as distinct entities or things. We can speak of the quantity of sound caused by the explosion of a cannon or by the blowing of a penny whistle; the quantity of light emitted by a gas-jet or by a farthing rushlight; the quantity of heat required to melt a pailful of ice or to solder a metal joint, without implying by the term *quantity* a mass or volume of anything actually present. The term implies relative magnitude only. It is the answer to the question 'how much?' It implies the notion of more or less. When we speak of the magnitude of electricity present or passing, we speak of its *quantity*. When we read of the church spire destroyed, of trees riven to splinters, of wires fused, or of flocks killed, the damage done is due to the electricity passing, and the amount of that damage is proportional to its magnitude or *quantity*. If we take a piece of sealing-wax, a glass rod, or an ebonite comb, and rub it against the coat sleeve, we find it has the property of attracting feathers, straws, and other light bodies. Electricity has been excited upon its surface, and the force of attraction is found to increase with the quantity of electricity present. Conversely the force with which bodies are attracted is an indication of the quantity of electricity excited. Hence we learn that ELECTRIC QUANTITY *is the magnitude or amount of electricity present.*

3. We may also conclude that electricity has a physical magnitude, and like all physical magnitudes is capable of measurement and of reference to some standard. Since quantity implies the notion of more or less, we must be able to answer the question 'more or less than what?' Hence all physical magnitudes must have a standard of reference or *unit* with which comparison and therefore calculation can be made. The notion of more or less is implied by the number of these units which are present. If we wished to express in feet the distance between any two places, we

might say 'let the distance between A and B be '*f*,' or if we wished to express in gallons the volume of water in a tank, we might say 'let the capacity of the tank be '*g*,' *f* and *g* respectively representing '*f*,' feet, and '*g*,' gallons; *f* and *g* standing for any number whatever, and the *foot* and *gallon* being the units or standards of reference taken or understood. Again, if we wanted to find the quantity of electricity required to effect a certain purpose, we might commence by saying 'let the quantity of electricity be '*q*,' by which we should mean an unknown number, *q* units of electricity, and our investigation might bring it out ·01 of a unit, or 3 units, or 50 units, or any other number. The unit quantity of electricity in general use has been called a *weber*, from Weber, one of the great German philosophers. Thus we see that in the literal representation of a physical quantity we assume the existence of a standard or unit to which we give a name, as foot, gallon, weber, and we express indefinitely its numerical value by the letter used.

4. Whenever electricity has been produced by any means, the bodies which exhibit evidence of its presence are said to be *electrified* or *charged with electricity*, and their condition is said to be one of *electrification*. For instance, a cloud which discharges itself into the earth with a flash of lightning is said to be electrified. A piece of sealing-wax, when it has been excited by rubbing so as to exert attraction, is electrified. A glass rod, similarly treated, is electrified. The method of electrification in the two latter cases may be precisely similar, yet the character of the electrification in each case is different. The sealing-wax and glass seem to be imbued with exactly opposite qualities: they attract neutral bodies; but repel each other; they are the antitheses to each other, like heat and cold, light and darkness. By an arbitrary convention the electricity excited on glass has been called *positive*, while that excited on sealing-wax has been called *negative*. All electrified bodies are either positively or negatively electrified. A thundercloud may at one time be positive and at

another time negative. When a cloud charged positively approaches a cloud equally charged negatively, discharge takes place between them, and complete neutrality, or zero, results. This justifies the use of the opposite terms.

5. Whenever we walk upstairs or ascend a hill we are conscious of having exerted ourselves. We have in fact raised our bodies through a certain height against the influence of the force of gravity. We have done work upon our bodies; and whenever we make an effort against a force of any kind, through any distance, we *do work*. Thus a horse does work in drawing a load, heat does work in converting water into steam, and thereby driving trains and propelling vessels. Electricity does work when it moves substances against the force of gravity, or when it flows against resistance, and indeed electrical phenomena are indicated by work done, obstacles overcome, or force exerted.

6. An electrified body acquires a certain quality or condition by which it possesses this power of doing work. In the same way that a poker placed in the fire acquires a high temperature before it burns the hand, or as water acquires a high pressure before it bursts the pipe, so electricity acquires a certain condition before it is capable of doing work. The property possessed by electricity, which is analogous to temperature and pressure, is called *potential.*

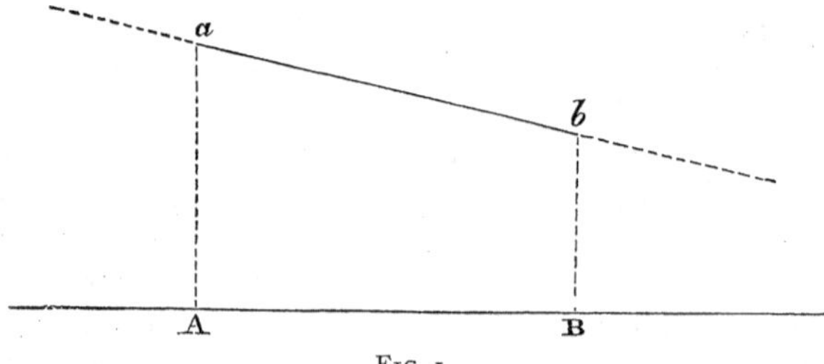

Fig. 1.

If it be desired to transfer heat from A to B, it is essential that the temperature at B be lower than that at A; and if it be desired to cause a flow of liquid or gas between two such points, it is equally essential that there be a difference of pressure between them. So if we desire to transfer electricity from A to B either along a conducting wire, such as that of a submarine cable, or through the air, it is imperative that the potential at B be less than that at A.

Hence, POTENTIAL *implies that function of electricity which determines its motion from one point to another.*

And the difference of potential, which determines the amount of this motion, is called *electro-motive force.*

7. The transference of electricity, such as that from a charged cloud to the earth, from a rubbed glass to a rubbed comb, a signal from Europe to America, may take place in different *times*; the path between A and B offers obstruction to the passage of electricity; the medium through which it passes, whether composed of an air space, or of any conducting material, is an obstacle to be overcome. Electricity in motion does work, decomposes liquids into their constituent elements, generates heat, produces magnetism, &c.; and the amount of work done with the same electro-motive force depends upon the resistance to be overcome; hence the term RESISTANCE *implies that quality of a conductor in virtue of which it prevents more than a certain amount of work being done in a given time by a given electro-motive force.*

The unit (§ 3) of resistance is called the *Ohm*, from Ohm, the German physicist, who determined mathematically the laws that regulate the flow of electricity. It is convenient for brevity's sake to use a symbol to represent the ohm as we use ° to represent degrees, and ′ minutes. The symbol used by us is ω, the Greek *omega.* Thus we say that the resistance of a wire between London and Birmingham is 1500^{ω}, and of one of the Atlantic cables is 7220^{ω}.

8. Those bodies which offer very great resistance to the passage of electricity through them are called *insulators*; those which offer very little resistance are called *conductors.*

The difference between a conductor and an insulator is one of degree only. Thus the resistance of a given volume or mass of metal or water is very small, while that of glass, ebonite, or air, is very great. The property of matter which determines its resistance is evidently molecular, for it varies with and is dependent upon the mass of the body as well as the character of its physical structure and condition. For instance, water, when a liquid, is a conductor; when a solid, an insulator; while many substances when cold are insulators, and when hot conductors.

9. The transference of electricity from one point to another is called a *current.* Whether it be a flash of lightning, or a simple spark between a piece of rubbed glass and the finger, or a signal sent from Europe to America, it is a transference of electricity. *To produce a current we must have two points at different potentials separated from each other by a resisting medium.* To produce a continuous current these points must be maintained at different potentials. A current will flow from the higher potential to the lower potential so long as a difference of potential continues, but when the potentials are equalized it ceases.

Hence we see that what is understood by the term CURRENT *is an apparent transference of electricity from one point to another to produce equalization of potential.* And one current differs from another only in its *strength*—or, in other words, in the quantity of electricity which is transferred by each in the same time. A current always is supposed to flow from the point of higher potential to that of lower potential. The former point is taken to be positive to the latter point; and, *vice versâ*, the lower point is taken to be negative to the higher point. The terms positive and negative currents are frequently used, but they are misnomers. There is only one current flowing from A and B (fig. 1), and it varies in direction. It is quite correct to apply the term positive or negative to currents *with respect to a given point*, and by those terms to imply direction only, for while stationed at a given place currents may flow *from* or *towards* us. But it is quite incorrect to speak of positive or negative currents without reference to a given point, for what is a positive current at one point is a negative current at another. The current is, however, supposed to flow from the body positively charged to that negatively charged. For the sake of convenience the potential of the earth is always assumed to be *zero* ; so that when we speak of the potential of a body, we really speak of the difference between its potential and that of the earth. In the same way the freezing point of water is taken to be the zero for temperature. This by no means assumes that the earth has *no* potential, for every thunderstorm and every telegraph line tell us that it has, and we shall have to speak of phenomena which show us that different portions of the earth's surface have different potentials at different times.

10. A current can only be *constant* when we have two points separated from each other by an invariable resistance, and *maintained* at the same difference of potential. The material conveying the electricity, whether it be earth, air, water, or matter in any form, separately or conjointly, is called a *circuit*, and *the circuit is the whole path along which the electricity is supposed to flow.*

11. These are the principal terms, independent of all hypotheses, which are used in the science of electricity in its application to telegraphic purposes, and it is upon their clear comprehension that the ease or difficulty of the mastery of the technical details of telegraphy depends. The nature of electricity itself is not known, nor is it necessary to the telegraphist that it should be known by him. He is only interested in its quantitative measurement and its application to practical purposes. Let him master its elementary principles, its general ideas, its properties and its conditions, and he can well afford to leave to physicists the discussion of its nature, and to mathematicians the determination of its laws.

CHAPTER II.

THE BATTERY.

12. If two plates of different metals, say pure zinc and platinum, be immersed in any acidulated solution, say sulphuric acid and water, then so long as the two metals are kept apart no action whatever is observed to take place between them. But immediately they are metallically connected together, whether by being brought into immediate contact at any point or by means of a wire, and so long as they remain so, the zinc is chemically attacked and eaten away; the acidulated water is decomposed into its constituent elements, one of which unites with the zinc, forming a salt of that metal, and bubbles of the other—hydrogen gas—are seen to form upon and to rise from the platinum plate.

If a wire has been employed to connect the two plates it does not remain quiescent, but in various ways it gives evidence of molecular disturbance and of the power with which it is imbued.

a. If it be dipped in a mass of iron filings, the little particles of iron will cluster around and apparently adhere to it.

b. If it be wound round a piece of soft iron it will render that iron for the time being magnetic, and almost immediately it is removed the iron loses all trace of magnetism.

c. If it be placed in the immediate neighbourhood of a freely suspended magnetic needle, the needle will at once exhibit a tendency to place itself at right angles to it.

d. If the wire be broken and the ends immersed in water, the water will be decomposed; oxygen collects at the end of the section proceeding from the platinum plate, and in this *nascent* state forms as a rule an oxide of the metal composing the wire. Hydrogen arises from the end of the section coming from the zinc plate.

The other manifestations of the power which the wire possesses, viz. its heating properties, its production of light, and its physiological effects, may be passed over, for they are not employed in any way in practical telegraphy.

13. To this combination of two different metals in acidulated water the name of a *cell* is given; and a series of such cells, properly arranged, forms the *galvanic or voltaic battery.* It is convenient to represent a cell symbolically or conventionally by a thick and thin line of different lengths—the former representing the zinc and the latter the platinum plate, as in fig. 2. A battery is similarly represented by a combination of these, as shown in the same figure.

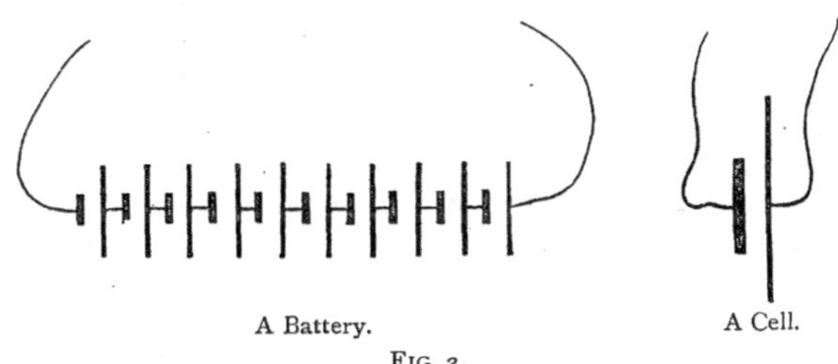

A Battery. A Cell.

FIG. 2.

14. The action observed is said to be due to a current which is conventionally assumed to start from the zinc plate, to pass through the liquid to the platinum, and thence to return by means of the wire to its starting-point. This term 'current' (§ 9) is purely a convention of language, and must not be taken to imply in any way the actual transference of matter from one point to another. The word was introduced in the early days of electricity, when electricity was believed to be a fluid, and it has ever since been retained. The power which the wire possesses in virtue of this transference of electricity, or, as we may now call it, the *strength* of the current, varies according to the metals

which are employed in the cell, as well as to the solution in which they are placed. In water acidulated with sulphuric or nitric acid the maximum effect is obtained when the metals farthest apart in the following list are combined:—Silver, copper, antimony, bismuth, nickel, iron, lead, tin, cadmium, zinc.

In water acidulated with hydrochloric acid the above order is modified as follows:—Antimony, silver, nickel, bismuth, copper, iron, lead, tin, cadmium, zinc.

15. Various theories have been advanced to account for the determination of the difference of potential resulting in this peculiar action. Volta was of opinion that it originated simply and wholly from the contact of the two dissimilar metals, and in this view he was supported (amongst others) by Ritter, Pfaff, Ohm, and Biot.

Faraday, on the other hand, maintained with Fabroni, who suggested the theory in 1792, that the prime cause was chemical action. He performed an enormous number of experiments in order to verify this opinion, and with the apparatus which he then had at his disposal they seemed to be convincing. This theory, known as the *chemical theory*, as opposed to Volta's *contact theory*, numbered amongst its supporters Wollaston, Davy, De la Rive, Daniell, and many others.

Now, however, by means of the delicate apparatus invented since Faraday's time, it has been distinctly shown that the mere contact of two dissimilar metals does determine a difference of potential between them, and thereby (§ 9) gives the prime conditions of a current. Based upon this a theory has been advanced which goes far to unite the rival contact and chemical theories, and which has been adopted by Helmholtz, Sir William Thomson, and several of the leading physicists of the present day. It is this. When two dissimilar bodies are brought into contact a difference of potential is determined between them. In figs. 3 and 4 let A represent a zinc plate, B a copper plate, both plunged in acidulated water, and CC′ a connecting wire of copper, which is shown in fig. 3 as divided into two sections, CA and

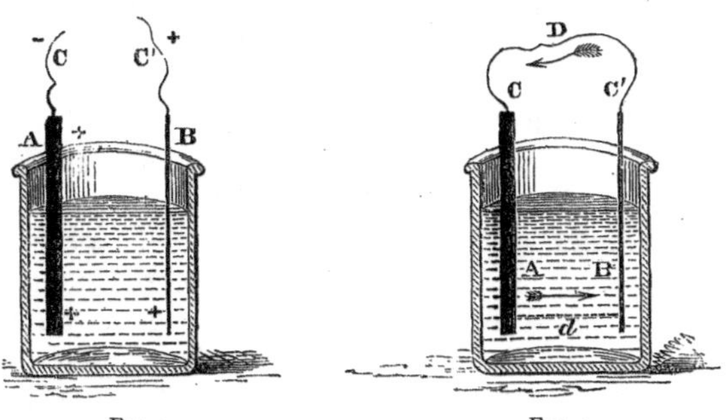

FIG. 3. FIG. 4.

C′B. Taking fig. 3, CA being placed in metallic contact with A, the latter has a positive potential at once determined to it relative to that of CA, and CA is therefore negative to A, B remaining in a neutral condition. A and B, which are separated from each other by a liquid, are therefore at first in different electrical conditions, but the conducting liquid at once reduces them to the same potential, and in doing so it is decomposed; consequently A and B are now in the same electrical condition. C′B acquires the same potential as B, because it is the same metal, and is in metallic connection with it; it has therefore a positive potential relative to CA. Thus the points C and C′ are at different potentials. Now suppose (fig. 4) C and C′ to be united; a current must flow because these two points are at different potentials. C retains the same potential relative to A, but the wire CC′ tends to reduce B to the same potential as C. The equilibrium is disturbed between A and B. The potential of B falls below that of A, a current flows, the liquid is decomposed, and this decomposition of the liquid exercises a counteracting influence; it endeavours to keep B at the same potential as A. The consequence is that B assumes a potential between that of A and that of C. Similarly any point D in CC′ has a potential between that of C and that of C′, and any point *d* in the liquid between A and B has a potential between that of A and B. Thus we have a constant fall of potential from the zinc through the liquid back to the zinc again through the connecting wire.

The result of all this is that what we may call a continuous flow of electricity is kept passing through the liquid from A to B, tending to keep the potential of the latter up to that of the former; and a continuous flow of electricity from C′ to C through CC′, tending to reduce the potential of B to that of C. A current therefore is said to pass in the direction A B C C′ when the cell is at work. The first flow is determined by the difference of potential due to contact, and the continued flow is maintained by the chemical decomposition of the liquid.

16. The zinc is named the positive plate or element, the copper the negative plate or element. These terms positive and negative convey no meaning of themselves; they are merely intended to denote the antagonism which exists between the two elements. The liquid itself is decomposed; the hydrogen (H) rising at the copper (Cu) plate leaves it untouched; the oxygen (O), however, attacks the zinc (Zn) plate, and gradually eats it away. Assuming first that pure water (H_2O) is made use of, the action which takes place may be symbolically represented thus :—

Before contact—

$$Cu,\ H_2O,\ H_2O,\ H_2O,\ Zn.$$

After contact—

$$\overline{CuH_2},\ \overline{OH_2},\ \overline{OH_2},\ \overline{ZnO}.$$

But zinc oxide (ZnO) is insoluble in water, and if pure water were used the action would at once cease, because the zinc plate becomes covered with an insulating compound. Hence sulphuric acid is added (H_2SO_4), which replaces the zinc oxide by the soluble zinc sulphate ($ZnSO_4$), leaving the zinc plate clear for further action. This action is symbolically represented thus :—

Before contact—

$$Cu,\ H_2SO_4,\ H_2SO_4,\ H_2SO_4,\ Zn.$$

After contact—

$$\overline{CuH_2},\ \overline{SO_4H_2},\ \overline{SO_4H_2},\ \overline{SO_4Zn}.$$

17. Although the deposition of hydrogen upon the copper plate is quite harmless so far as the copper itself is concerned, yet it has a very deleterious effect upon the general working of the cell. The working is impeded, and the strength of the current is very sensibly diminished by it. To this obstructive action the name of *Galvanic Polarisation* has been given.

It is due to the fact that the free hydrogen accumulating upon the copper plate behaves with respect to it in a manner almost exactly similar to that of the zinc itself—that is to say, the hydrogen assumes a positive potential relative to the copper. The result is very nearly the same as though two plates of zinc were opposed to each other. Should this be the case, no difference of potential could be determined, and consequently no current would be obtained.

18. Another injurious effect which the hydrogen ultimately exercises upon the action of the battery is due to the facility which it possesses for reducing the metals from their salts. The zinc sulphate which in time accumulates by the action of the cell is reduced by the hydrogen as soon as they come into contact with each other; the zinc is thrown down upon the copper plate, and therefore zinc is eventually opposed to zinc, and then the current entirely ceases to flow.

19. It will thus be seen how essentially important a matter it becomes to prevent the presence of free hydrogen upon the negative plate.

This object has been attained in various ways.

In Smee's battery the deposition of hydrogen on the negative plate was prevented by mechanical means. He coated the plate with finely-divided platinum, and the hydrogen, being readily discharged from its roughened surface, rose in bubbles to the surface of the liquid. This battery is not practically employed in telegraphy now, and may therefore be passed over without further comment.

20. Daniell invented the battery which bears his name. Taken all in all it may be regarded as the most perfect now in practical use for telegraphic work. It appears under various modifications, the principle, which is as follows, remaining the same throughout :—

Zinc and copper are employed as the positive and negative plates, but instead of being in the same liquid they are placed in different liquids, which are separated from each other by a porous partition. The liquid surrounding the zinc is diluted sulphuric acid ; that surrounding the copper a solution of the copper sulphate ($CuSO_4$). The part played by the latter is the distinguishing feature of Daniell's battery. The instant the two plates are united with each other action commences ; the zinc plate is attacked, and a salt of that metal is formed ; the hydrogen liberated at the copper plate reduces the copper sulphate, expelling from it the metallic copper, which is thrown down in a perfectly pure state upon the copper plate of the cell. The hydrogen then combining with the molecule SO_4 forms sulphuric acid (H_2SO_4), which, finding its way through the porous partition into the zinc cell, maintains the solution there at a constant strength.

The consequence of this is that the positive plate is gradually eaten away, and the liquid surrounding it becomes a solution of the zinc sulphate ; the copper sulphate is reduced, but the negative plate—the main point to be looked after—is kept perfectly clean and bright by the deposition upon it of pure metallic copper thrown down by the hydrogen from the solution of copper sulphate.

The action of a Daniell's battery may be symbolically represented thus :—

Before contact—

$$Zn \quad SO_4H_2, \; SO_4H_2, \, || \, SO_4Cu, \; SO_4Cu, \; Cu.$$

After contact—

$$\overline{ZnSO_4}, \; \overline{H_2SO_4}, \; \overline{H_2SO_4}, \, || \, \overline{CuSO_4}, \; \overline{CuCu}.$$

21. There are various forms of Daniell's battery in use. One of the most convenient, and that which is generally employed in England, is the following :—

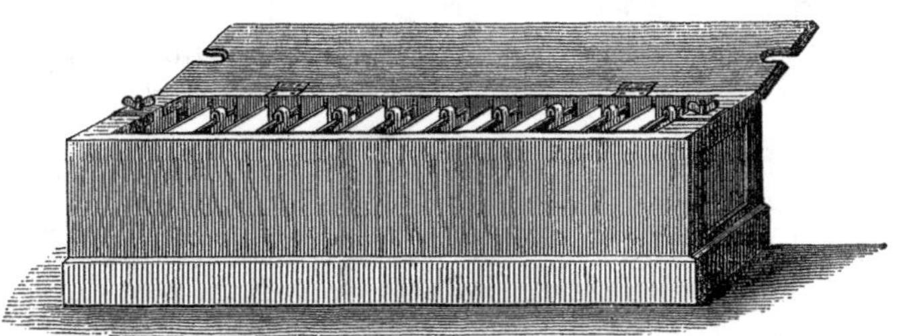

FIG. 5. $\frac{1}{10}$th real size.

A teak trough (teak is selected on account of its durability, and from the fact that it shows little tendency to warp), measuring two feet in length, six inches in width, and five and a half inches in depth, is divided into ten spaces or cells, which are separated from each other by slate partitions. It is then coated throughout, including the slate partitions, with marine glue.[1] The object of this is to render the trough perfectly water-tight and prevent any leakage from one cell to another.

When the trough has been thus served each cell is subdivided by a porous partition, composed of unglazed porcelain, of a uniform thickness of rather less than a quarter of an inch throughout.

[1] The marine glue, patented by Jeffrey in 1842, is formed by dissolving one pound of caoutchouc in four gallons of naphtha, and allowing this to stand for ten or twelve days. Two parts of shellac are then added to one part of this mixture, and the compound thus obtained is cooled on marble slabs.

The zinc plate (fig. 6), measuring $3\frac{1}{2}$ in. × 2 in., is suspended into one of the divisions thus formed, which is filled up to within a quarter of an inch of the top of the plate with pure water; the copper plate, which is 3 in. square, is placed in the other, amongst crystals of copper sulphate; this division is then filled up to within about the same distance of the top of the copper plate with pure water. The copper plate of one cell is connected with the zinc of the succeeding cell by a copper strap passing over the slate partition. The plates in fact, before being fitted into the trough, are permanently connected with each other, and are issued in pairs. One end of a copper strap measuring $2\frac{1}{2}$ in. long by $\frac{3}{4}$ in. wide is fastened to the copper plate by means of a copper rivet, and the other end of it, after being well tinned for about $\frac{3}{4}$ in., so as to insure good metallic connection, is placed in a closed mould into which molten zinc is run.

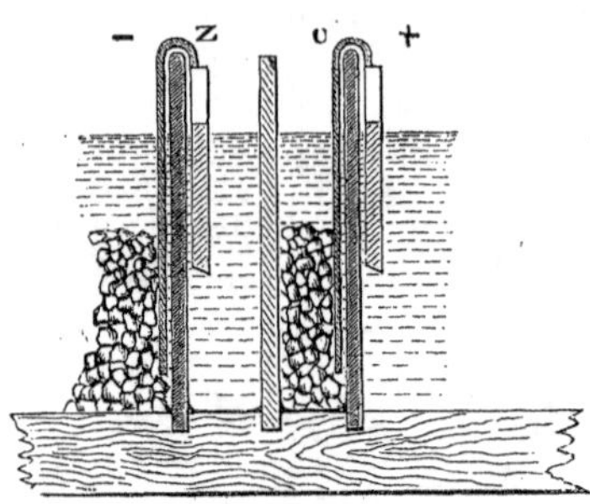

FIG. 6. $\frac{1}{4}$th real size.

The last copper and the last zinc are each connected to a brass binding screw.

22. In setting up a Daniell's battery, there are various points to which special attention must be given. A disregard of any one of them will more or less mar its action.

a. The copper sulphate, which is manufactured by dissolving scales of cupric oxide in sulphuric acid, must be of the purest possible description. The foreign ingredient mainly to be found in it is iron, whose presence may be ascertained by the following test:—The copper sulphate, like all the copper salts, forms with excess of ammonia a deep blue solution, whilst the iron sulphate, under similar circumstances, is precipitated as a dirty-brown powder. If, therefore, to a solution of copper sulphate ammonia be added until this deep blue colour is obtained, the amount of iron present, provided there really is any, can be readily known. In good copper sulphate it should never exceed ·55 per cent.

b. The metals, but more especially the zinc, should be as pure as can possibly be obtained. This applies to the metals, not only of Daniell's battery, but of every other species of battery as well. For if any foreign ingredients make their appearance, the action of the battery is seriously interfered with; the effect is the same as though a number of small plates of different metals were opposed to each other. Local currents are generated; the plates are needlessly wasted, and the general strength of the battery is impaired.

Let fig. 7 represent a portion of a zinc plate containing several particles of iron, tin, or lead, which are the usual impurities to be met with in it. The contact of either of these (say lead) with zinc determines to the latter a positive potential; a liquid arc intervenes (fig. 7 A), and all the required conditions being present, a current starts from the zinc to the lead through the liquid. Owing to the *local action* which is thus commenced, the zinc plate is eaten away to no purpose, the liquid is decomposed, and the hydrogen which is liberated partially polarizes the zinc plate (§ 17). The consequence is that the *resultant*[1] current is materially weakened, and in time would be wholly stopped.

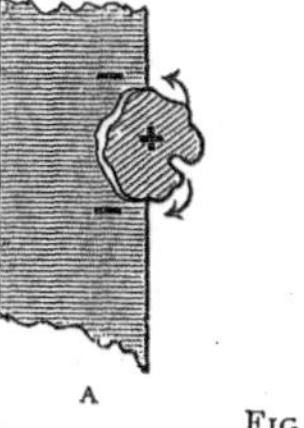

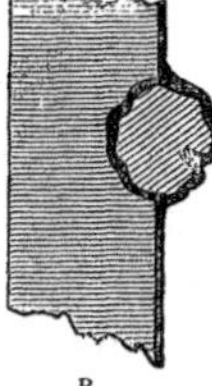

FIG. 7.

The possibility of anything like this occurring is pre-

[1] The term *resultant* implies the ultimate effect of a series of actions which may be similar or opposite in their character. There may be several causes present to determine currents in the same or different directions, and the resultant current is the final result or algebraic sum of all those causes.

vented in various ways. Pure zinc, on account of its being so expensive a metal, cannot be employed. The object, however, is attained either by covering the zinc with mercury, a process called *amalgamation*, or by employing a solution of zinc sulphate in place of acidulated water in the zinc cells.

Mercury possesses the power of combining with several of the other metals, and forming alloys which are known as amalgams. Zinc may be amalgamated by being first cleaned with hydrochloric or sulphuric acid, and then rubbed over with mercury. The liquid arc can then no longer intervene between the various impurities in the plate ; a mercurial metallic arc is substituted in its place, fig. 7 B. Consequently the conditions for a local current are destroyed, and no 'local action' on the surface of the plate can take place ; at the same time a perfectly homogeneous surface is presented for the general working of the battery.

This course is not, however, adopted in the Daniell's battery employed in telegraphy ; it has been found more advantageous in every respect to adopt the suggestion first made by Mr. Fuller, viz. to employ a solution of the zinc sulphate if the battery is to be brought at once into use. But it will be seen (§ 20) that this salt, the zinc sulphate, is spontaneously formed in the action of the battery. Consequently if the action is allowed to go on for some time, say forty-eight hours, before the battery is actually required, it becomes unnecessary to use at the outset anything more than water in the zinc cell.

c. The copper sulphate must be used in the form of crystals and not as a powder. In the latter state it dissolves slowly, and in time adheres so tenaciously to the cell that it can with difficulty be removed.

d. Care must be taken that the zinc plate does not touch the porous partition. Should it do so a local action commences at once, due to the fact that the sulphuric acid contained in the copper sulphate (§ 20), which in time makes its way through the porous partition, has a far greater affinity for zinc than copper. It consequently leaves the latter, which is precipitated in a metallic state on the side of the porous partition, and immediately gives rise to a local action.

23. Batteries such as those described, in which these precautions have been taken, will remain in constant action for a month without requiring any attention whatever. At the expiration of a month it becomes necessary to refresh them, and the following points must then be seen to.

a. The solution in the zinc cell should not be supersaturated with the zinc sulphate. The result of this would be the deposition of crystals on the zinc plate, the copper strap, and along the edges of the cell, whereby the liquid is carried off by capillary action, and short circuits are formed between the cells. Should this occur, a portion of the liquid must be drawn off, the cell filled up with water, and the crystals removed. The solution is in the best possible state when it is semi-saturated with the zinc sulphate; its conducting power is then at a maximum.

b. The zinc plate should be examined, and if there be any quantity of what at first sight appears to be black mud upon it, this should be scraped off and carefully laid aside. The 'black mud' contains the purest copper, and its presence on the zinc plate is thus accounted for:—Liquids differing in specific gravity and separated from each other, either by gravity alone or by a porous diaphragm, possess the power of gradually diffusing into each other, and in time forming one mechanical solution. The specific gravities of a solution of the zinc sulphate and of a solution of the copper sulphate vary. They consequently mingle with each other in course of time through the porous partition; but no sooner does the copper sulphate enter the zinc cell than the sulphuric acid leaves it and unites with the zinc, for which, as has been already observed (§ 23) it exhibits a decided preference. The copper of the copper sulphate is thus set free and deposited on the zinc plate. The action of the battery is thereby gradually weakened, until eventually, when the zinc plate is covered with copper, the current entirely ceases to flow. It is just in effect as if two plates of the same metal were employed, between which no difference of potential can of course be determined. The copper, on account of the finely-divided state in which it is precipitated from its sulphate, loses its bright metallic lustre and speedily becomes oxidised.

c. The copper cell should be examined, and if the crystals of copper sulphate have been nearly exhausted a fresh supply should be added, and water poured in to supply the place of that which may have been carried off by evaporation.

d. Special care should be taken that the connecting straps and the terminal binding screws are kept bright and clean.

24. The battery, at the end of two months, if it has been in constant use, and three months if it has been but moderately worked, should be thoroughly cleaned throughout. The solution in the zinc cell is first drawn off by means of a syringe and placed for further use in a vessel, into which it is advisable to throw a few scraps of zinc; for any copper which may be held in solution will thus be thrown down and only the zinc sulphate remain. The liquid in the copper cell is drawn off in the same manner, any crystals which may remain being taken out. The plates are next removed, well scraped and cleaned, the 'mud' obtained from the zinc being carefully preserved in a box provided for the purpose.

The porous partitions are then cleared of the copper with which they have become partially encrusted. The presence of copper on them cannot well be prevented; it is one of the results of a local action which owes its origin partly to the impurities contained in the zinc plate which have not been effectually got rid of, and partly to those which, in the shape of metallic dust or small pieces of carbon, are occasionally to be met with in the porous partition itself. It may be prevented to some extent by saturating the bottom of the porous diaphragm to the height of about half an inch with melted wax or paraffin. The cells must be well rinsed out, the metal deposited in them scraped off, and every particle of foreign matter removed.

25. In recharging the battery the liquid drawn off from the zinc cells is again employed in them, having been previously diluted with water if necessary; the plates, after being well cleaned, are inserted as before, such as were defective having been replaced by others. The copper cell is then filled up with the copper sulphate, as was done in the first instance.

26. It is essential that the battery should be placed in a dry position, free alike from the extremes of heat and cold, and be protected as far as possible against the accumulation of dust upon it. If it rests on damp ground the strength of the current obtained is weakened; for the damp ground being a conductor tends to abstract a certain portion of the current.

The batteries are usually placed upon wooden racks, the boards of which should be of a triangular section, as shown in fig. 8.

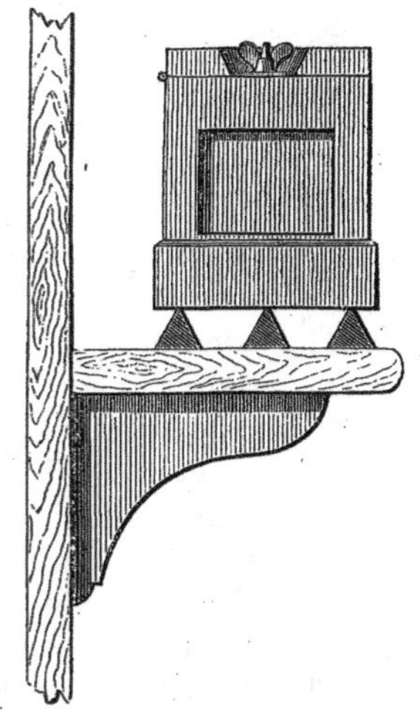

FIG. 8. ⅛th real size.

27. Various forms and sizes of Daniell's battery are used, depending chiefly upon the work which they have to do. One, which is employed to a considerable extent in Great Britain, is the Chamber form of Daniell's battery, introduced by Mr. Muirhead. Into a vessel of glazed porcelain or of ebonite a porous earthenware pot, of the shape shown in fig. 9, and dipped for about

half an inch in melted paraffin, is placed. The latter contains the copper plate with the solution and crystals of the copper sulphate; the zinc is placed in the porcelain vessel, which is filled up as before with water. The action as well as the mode of treatment is exactly similar to that which has been already described. At a large station, where every facility is afforded for cleaning batteries, the 'Chamber' cells can be more easily handled than those of the trough; whilst if it be found necessary to employ increased battery power, a few of these cells can be very conveniently added to those previously in existence.[1]

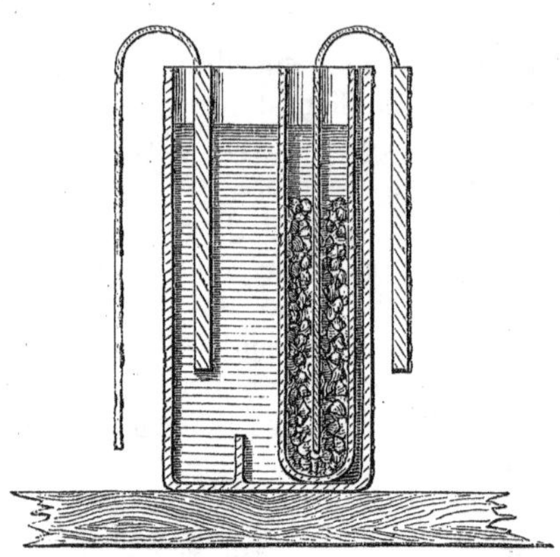

FIG. 9. ⅓rd real size.

28. The conditions to be fulfilled by a good working battery for ordinary practical purposes are :

1st. That the strength of the current obtained from it should be constant.

2nd. That the materials used in the construction and maintenance of it should not be expensive.

3rd. That when the battery is not being worked, there should be no waste of the materials employed.

29. The Daniell's battery fulfils the first condition as satisfactorily as any battery which has hitherto been invented.

[1] Over 20,000 of these cells are employed in the Central Telegraphic Station, G. P. O., London. They are ranged in series on movable longitudinal racks, which are protected by hinged covers. Troughs are thus dispensed with.

Local action, causing a variation in the strength of the current, does, as has been pointed out (§ 22), take place ; but if the precautions indicated above to prevent this are taken, the variation is so slight as to be imperceptible in practical working, and no inconvenience is felt from it.

In point of cheapness both in construction and maintenance, the Daniell contrasts favourably with its rivals. A battery similar to that which has been described above, charged and ready for use, costs 1*l.* ; in the course of twelve months if the battery has been fairly worked, ten pounds of sulphate of copper are used ; and apart from the labour of refreshing and cleaning out, the annual cost of the maintenance of it may be set down at 7*s.* 6*d.*

It does not fulfil the third condition. Even when Daniell's battery is at rest there is a waste of the materials employed. By reason of the action, to which reference has been made (§ 23), the liquids diffuse into each other through the porous cell, and the copper sulphate is gradually reduced.

On account of this the porous cells of a Daniell's battery which is required only for occasional use, are made considerably thicker than those already described ; in this way the mixture of the two solutions is retarded, but at the same time the resistance to the current as it passes from plate to plate is increased. Several modifications of the Daniell's battery, to which reference will be made hereafter, have been constructed, with a view to prevent as far as possible the diffusion of the liquids.

30. Next to Daniell's, the battery which of late years has obtained most favour in Great Britain is that invented by M. Leclanché, of the Eastern Railway of France.

Zinc is employed by him as the positive element, and binoxide of manganese (MnO_2), the pyrolusite of mineralogists, as the negative element. This mineral is to be found in Germany, France, Hungary, Brazil, Cornwall and Devon, &c., and is one of the main supplies of oxygen. For use in the battery it is broken up into coarse grains and carefully sifted; in this way all that exists in the form of a powder is got rid of. It is mixed with an equal volume of carbon crushed to about the same state as the binoxide of manganese itself. An earthenware porous pot, into which a plate of carbon has been placed, is then filled with this mixture. The zinc, which is in the shape of a rod, is surrounded by a solution of the chloride of ammonium (NH_4Cl), the ordinary sal-ammoniac; and when it is connected with the carbon plate, the following action takes place :—The zinc is attacked by the chlorine; chloride of zinc is formed, and dissolved in the liquid. The other constituent of the sal-ammoniac besides chlorine, namely ammonium (NH_4) is immediately, on being set free, oxidised by the peroxide of manganese, and ammonia and water are thereby formed. So long as this simple action goes on unimpeded by any other, galvanic polarization is prevented, and the strength of current obtained from this combination remains constant. The binoxide of manganese is reduced to a lower oxide known as the sesquioxide (Mn_2O_3). What actually takes place may be symbolically represented as follows :—

Before contact—

$$Zn,\ 2(NH_4Cl),\ |\ 2(MnO_2), C.$$

After contact—

$$\overline{Zn\,Cl_2},\ \overline{2NH_3},\ \overline{H_2O},\ |\ \overline{Mn_2O_3},\ C.$$

The results of the action are, the formation of chloride of zinc, free ammonia, water, and the reduction of the binoxide of manganese to the sesquioxide. Leclanché has brought out three sizes of cells; that which is generally adopted is of the form shown in fig. 10. Into a glass vessel containing a solution of sal-ammoniac a zinc rod is placed; the porous pot containing the carbon plate and the mixture of pounded carbon and binoxide of manganese is next inserted into it. This carbon plate is fitted with a lead top, into which a binding screw is fixed for the purpose of connecting it with the wire proceeding from the neighbouring zinc. Lead is employed in preference to any other metal chiefly on account of its stability, and it is of great importance that good contact should be insured between it and the carbon.

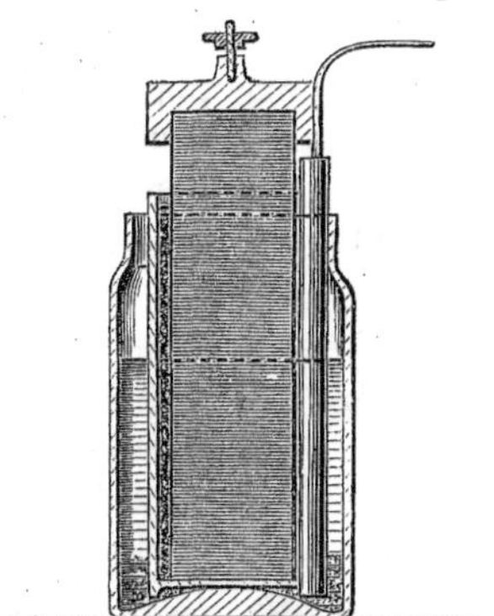

FIG. 10. $\frac{1}{4}$th real size.

31. In setting up a Leclanché's battery the following points must be carefully attended to :

a. A strong solution of sal-ammoniac should be used ; for although the chloride of zinc is an easily-soluble salt, yet in the action of the battery double salts—oxychlorides of zinc and zinc-ammonic-chlorides—are formed, which require a more concentrated solution of sal-ammoniac, and longer time to dissolve, than does the simple chloride of zinc.

The formation of these double salts of zinc is the result of a secondary action which, after the battery has been kept steadily at work for a short time, makes its appearance and seriously interferes with its constancy. So long as chlorine only is set free at the positive plate and ammonium liberated at the negative, so long is galvanic polarization averted ; but as soon as oxygen arises at the zinc and hydrogen unconsumed accumulates on the carbon—which actually does occur after continued working for a few minutes—that moment galvanic polarization ensues, a counter current is generated, and the resultant strength of current obtained from the battery is impaired. This galvanic polarization, along with every trace of a secondary action, speedily disappears if the battery be left to itself; it is not observable at all if the battery is called into play only at intervals.

b. The porous cell should stand only half its height in

the solution, as the drier its contents are the better for constant working.

c. The connecting wires from the carbons to the succeeding zincs must be carefully protected. This is done by covering them with paint, tar, gutta-percha, Chatterton's compound,[1] or any other substance of a similar nature. India-rubber has been found to answer the purpose as well as anything. The object of this is to prevent the free ammonia given off in the action of the battery from reaching the metallic wire; if the wire is exposed to the smallest extent, the ammonia attacks and gradually eats it through. The result is that the circuit

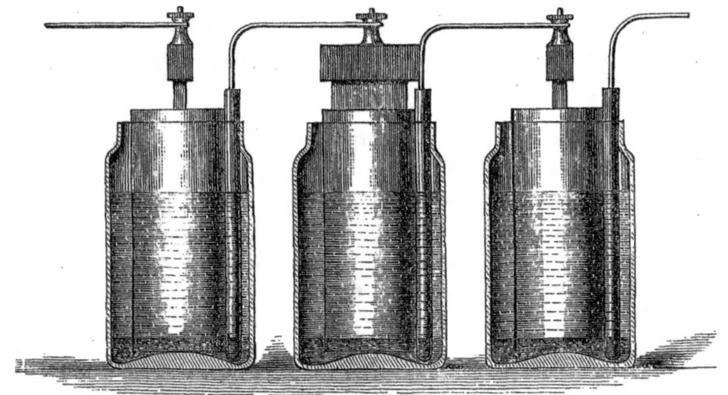

FIG. 11. ¼th real size.

is broken, and the battery is for the time rendered perfectly useless.

d. The binoxide of manganese which is used is of the form known as needle manganese. All the dust should be carefully removed from the coarse powder into which this

[1] Chatterton's compound is a mixture of resin, Stockholm tar, and gutta-percha, in the following proportions :—

1 pound of resin,
1 ,, ,, Stockholm tar,
3 pounds of gutta-percha.

is broken up. M. Leclanché found the presence of a small amount of fine powder in the porous pot to be not only injurious to the action of the battery, but also to interfere greatly with its constancy.

The top of the carbon pot is covered with marine glue if the battery is to be sent any distance. Care must be taken, however, to leave a hole in the marine glue so as to allow the air to escape when the pot is placed in the solution. The cells of a Leclanché's battery are joined up in the usual way to form a series. Fig. 11 shows how three of these cells are so connected.

32. A Leclanché's battery thus set up will remain in good condition for a period varying according to the amount of work which it is called upon to do. If it is required only for occasional use, such as the ringing of bells for either signalling or domestic purposes; or, if it is employed upon a speaking circuit along which comparatively little traffic passes, it is really difficult to say how long the battery would last, provided the precaution is taken to replace every now and then a portion of the liquid in the zinc cell with pure water, adding at the same time a little sal-ammoniac, and if need be cleaning the zincs. Instances have occurred where it has been left for periods ranging from nine to eighteen months, or even longer, without being so much as looked at, and yet no complaint of its working was heard. On busy circuits, however, it cannot be relied upon to anything like the same extent as the Daniell. The zinc salts which are then formed do not admit of being readily dissolved by the solution of sal-ammoniac; the secondary action already alluded to makes itself felt; the strength of the current consequently varies, and constancy is lost. And not only this; the porous pots crack in considerable numbers; the glass cells occasionally break from no apparent external cause, and the connecting wires, if exposed to the slightest extent, are very liable to be eaten through by the free ammonia given off. A local action, too, is observed to take place between the iron con-

necting wire, the brass binding screw, and the lead top of the carbon plate. Salts of lead are there formed, causing disconnections in the circuit. An attempt has been made to get rid of this local action by welding or soldering the iron wire on to the lead top of the carbon plate, and issuing the elements in pairs, as is the case in the form of Daniell's battery, which has been described (§ 21). White lead is also speedily formed in considerable quantities at the junction of the carbon with the lead.

33. In the conditions to be fulfilled by a good working battery, Leclanché's battery possesses one decided advantage over Daniell's, and that is, that there is no waste of materials when the battery is not actually at work. For the diffusion which takes place in Daniell's battery cannot exist with the single fluid in Leclanché's. In point of cheapness, however, as well as constancy, the Daniell's battery holds its own. A ten-cell Leclanché, of the form described, would cost 2*l.* 2*s.*; the cost of maintenance, like the constancy, will vary according to the purpose for which it is employed.[1]

34. A form of Leclanché's battery which has recently been issued obviates, to a very great extent, many of the evils which are inherent to that described above. This new form is to all external appearance like the ordinary Daniell; a teak trough is similarly divided, and coated throughout with marine glue; the porous partitions are of the same form, and the only difference lies in the constituents of the cells. Into one a carbon plate, surrounded with the mixture of coke and binoxide of manganese, is placed; into the next a zinc plate is immersed in a solution of sal-ammoniac. The terminal zinc and terminal carbon are each connected to a brass binding screw; the intermediate plates are issued in pairs in exactly the same way as the plates of the Daniell; they are connected with each other by means of an iron

[1] Mr. C. V. Walker, F.R.S., has made a very careful examination of the cost of maintaining a Leclanché's battery, and finds it to be 15½*d.* per cell per annum, including stores, labour, and travelling.

wire, which is firmly welded into the zinc and the lead top of the carbon plate. This lead top is carefully covered with marine glue, and the connecting wire is likewise covered with a protective coating.

This form of battery thus dispenses entirely with the glass cells; the porous partitions in it are not so liable to crack as the pots in the previous form, and the short length of the connecting wire, especially if thoroughly well protected, reduces to a minimum any danger in connection with it. There can be no question that in every sense this is a decided improvement on the previous form; upon all, except the busiest circuits, it may be employed with safety and economy, for the maintenance now becomes an exceedingly simple matter; if the zincs are scraped clean and the solution of sal-ammoniac kept up—half being removed and replaced by water—say every three months, unless under exceptional circumstances, experience so far does not warrant our assigning a limit to the time that it will remain in action.

35. Other forms of constant batteries occasionally employed for practical purposes in England are Grove's and Bunsen's. A Grove's cell consists of a plate of zinc as the positive element, in dilute sulphuric acid, separated by means of a porous partition from a plate of platinum (Pt), the negative element, which is immersed in concentrated nitric acid (N_2O_5). When the circuit is completed the zinc is attacked, the soluble sulphate of zinc is formed, and the liberated hydrogen is oxidised to water by the concentrated nitric acid before it can settle on the platinum plate. The nitric acid is thereby reduced, and fumes of the peroxide of nitrogen arise in the form of a dark-brown vapour.

The action which takes place may be symbolically represented thus:—

Before contact—

$$\text{Zn } H_2SO_4 \mid N_2O_5 \text{ Pt.}$$

After contact—

$$\overline{ZnSO_4} \mid \overline{H_2O}, \overline{N_2}O_4 \text{ Pt.}$$

The expensive materials employed in this battery disqualify it for ordinary practical use ; it requires almost constant looking to, and its constancy, after it has been but a short time at work, and the nitric acid has got weakened, cannot be relied upon. The strength of current obtained from a Grove's battery upon short circuit is, compared with that from a Daniell's battery, roughly as 8 : 1 ; for this reason Grove's battery is largely employed for experimental purposes where a powerful current is required. It is admirably adapted for this, and was in fact originally designed specially for this purpose.

Bunsen's battery is similar to Grove's, with the exception of the negative element. The expensive platinum employed in Grove's battery is replaced in Bunsen's by carbon specially prepared for the purpose.

36. In France Leclanché's battery is largely employed, and has of late years supplanted to a great extent that which was formerly much in use, viz., the Marié-Davy. The principle of the Marié-Davy is very similar to that of the Daniell. The only difference is that a carbon plate in sulphate of mercury (Hg_2SO_4) is employed as the negative element in place of a copper plate in the sulphate of copper ; zinc is made use of as the positive element. There are two sulphates of mercury, the protosulphate ($HgSO_4$) and the bisulphate (Hg_2SO_4). The latter occurs in two modifications, viz. the yellow basic sulphate, which is insoluble, and the white crystalline acid or neutral sulphate, which is slightly soluble. In a solution of this white crystalline bisulphate the carbon plate is immersed. The zinc is placed at first in water, but owing to diffusion mercuric sulphate appears. By chemical affinity mercury is deposited on the zinc, which is thus self-amalgamated and sulphuric acid is set free.

When the circuit is completed the zinc sulphate is formed ; the liberated hydrogen throws down the mercury from its salt, and is thereby prevented from polarizing the carbon plate, which is quickly covered by the mercury instead. The sulphuric acid formed behaves as it did in the Daniell's battery (§ 20).

The action is as follows :—

Before contact—

$$Zn, H_2SO_4, H_2SO_4 \mid Hg_2SO_4, Hg_2SO_4, C.$$

After contact—

$$\overline{ZnSO_4}, \overline{H_2SO_4}, \overline{H_2SO_4} \mid \overline{Hg_2SO_4}\ HgC.$$

The strength of current obtained from this battery is considerably in excess of that from a Daniell's; the latter is about two-thirds of the former. So long also as the battery is not heavily worked the current is constant; but if used on a busy wire, the battery labours under the same disadvantage as Leclanché's. The hydrogen is given off in too large quantities to admit of its place being wholly taken by the mercury; the consequence is that some of the hydrogen adheres to the negative element; polarization of course ensues, and the constancy of the battery is interfered with. This combined with the expense incurred in its maintenance, on account of the high price of the mercury salts, prevents the battery from being employed to any great extent in telegraphy; it was tried in England, but had to be abandoned.

37. In Germany the batteries mainly in use are Meidinger's and Siemens and Halske's.

Both of these are modifications of Daniell's battery, and have the same object in view, viz., to prevent as far as possible the deposition of copper on the zinc plate by the diffusive action described in § 23, as well as to put a stop to the local action due to the impurities in the zinc.

Meidinger discarded the porous diaphragm altogether, and relied upon the effect of gravity for keeping apart the liquids employed in his form of battery. Previous, however, to going into this, it will be advisable to glance at the general principles of the so called 'Gravity' batteries.

It has been already mentioned that two liquids varying in specific gravity possess the power of diffusing into each other, and ultimately forming one mechanical solution. Graham showed that this process of diffusion was an extremely slow one, and Fick advanced the now universally accepted theory that the rate of diffusion among different liquids varies inversely as the square root of their specific gravities. Advantage has been taken of these facts in the arrangement of a galvanic battery, in which the porous partition is dispensed with altogether, and the liquids are kept apart by the force of gravity alone. A copper plate is

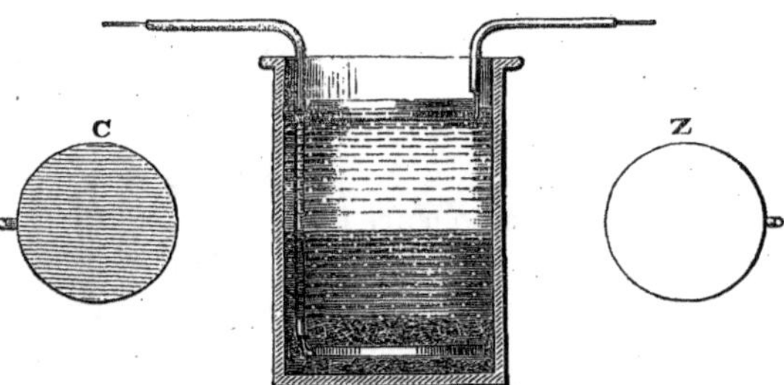

FIG. 12. $\frac{1}{8}$th real size.

placed at the bottom of the vessel (fig. 12), and over it a saturated solution of copper sulphate. A less dense solution of zinc sulphate in which the zinc plate is immersed is placed over this. The connecting wire leading to each succeeding cell is covered with india-rubber or gutta-percha, to protect it from the free acid formed in the action of the battery.

Such is the principle of all the gravity batteries; unless aided by some mechanical contrivance it has not proved a success. Absolute rest, so that the liquids may not be shaken up together, is indispensable for their working; and even when this condition is fulfilled, the waste of zinc and copper sulphate which takes place is far greater than in the case of the 'ordinary' Daniell's battery.

38. Sir William Thomson, availing himself of the fact that a *saturated* solution of zinc sulphate is considerably heavier than a *saturated* solution of copper sulphate (the specific gravity of the former being 1·44, that of the latter 1·18), has produced a form of gravity battery in which the order of the elements is the reverse of that described above. The zinc plate is immersed in a saturated solution of zinc sulphate, which is placed at the bottom of the vessel, and diminishes in strength upwards; over this is a saturated solution of copper sulphate, in which the copper plate is laid. The objections to this arrangement are that the zinc sulphate is found to gradually penetrate the entire liquid; in course of time the copper which is precipitated from its sulphate, falling on the zinc plate, entirely covers it, and when the zinc plate has to be removed for the purpose of being cleaned or replaced by another, this can be done only by disturbing the entire arrangement—that is to say, the solutions become mixed, and must be made up afresh before the cell can be restored to working order. By ingenious mechanical contrivances Sir William Thomson has, to a great extent, surmounted these difficulties, and brought out a battery whose use for some special kinds of work has been attended with excellent results, but which does not appear likely to be adopted as a practical telegraph battery for speaking purposes.

39. Returning now to Meidinger's element, it consists of a glass cell of the shape shown in fig. 13, containing a smaller glass vessel inside it and resting on its bottom. Into the latter a copper cylinder is placed; to this a copper wire, covered with gutta-percha, is attached, and thence passes to the next cell. Into the cell a zinc cylinder is inserted and supported upon the projecting edges. A wooden lid covers the whole, and through an aperture cut in its centre a glass vessel in the form of a test tube, with a few holes perforated in the bottom of it, is suspended, and reaches about half way down into the smaller vessel.

This element is set up by filling the test tube with crystals of the copper sulphate, and the vessel, up to within about a quarter of an inch of the top of the zinc plate, with a solution of the magnesia sulphate ($MgSO_4$. Epsom salts). The copper sulphate, after dissolving, passes through the holes in the test tube, and from its greater specific gravity settles at the bottom of the vessel. The zinc and magnesia sulphates, being specifically lighter, remain on the surface until, by diffusion, they gradually mingle with the copper sulphate; it then becomes necessary to re-charge the cell. In this, however, as in every other form of gravity battery, it is essential that the solutions should remain undisturbed, and every precaution be taken against their being shaken up and thereby mixed with each other.

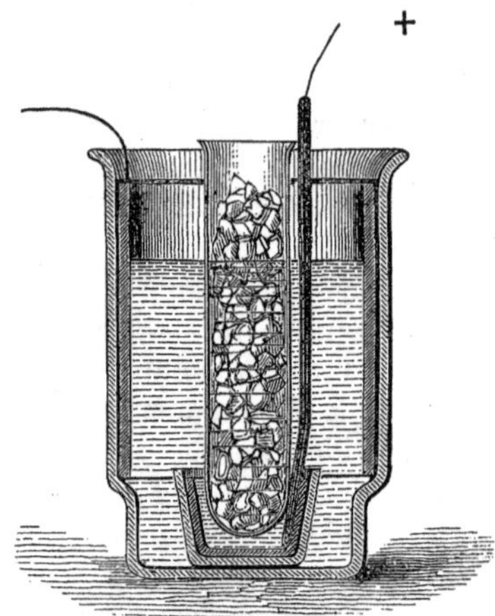

FIG. 13. $\frac{1}{6}$th real size.

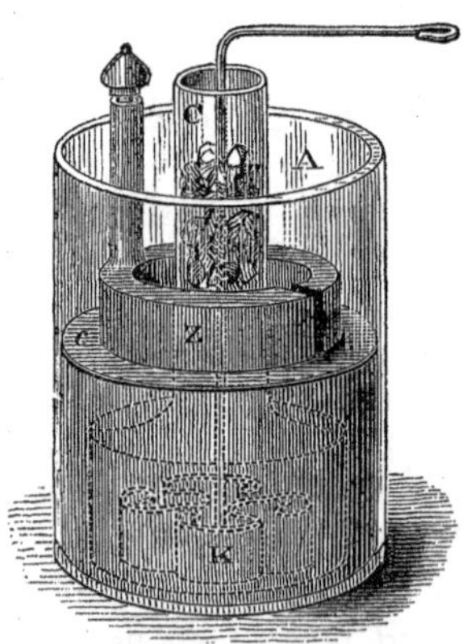

FIG. 14. $\frac{1}{6}$th real size.

40. In Siemens and Halske's form of Daniell's battery, the main point is the substitution of specially prepared paper pulp in place of the porous earthenware or unglazed porcelain partition of the ordinary form.

One of these elements is shown in fig. 14. A is a glass vessel, at the bottom of which a cross of sheet copper of the form K is placed; over this stands a tube, C, of unglazed porcelain, having its lower part widened out bell fashion. Into this tube a supply of crystals of the copper sulphate is placed and water filled in. The glass vessel is packed as far as *c* with the paper pulp, which in its preparation has been treated with sulphuric acid, and worked up into a homogeneous glutinous mass. This is well pressed down, and over it the zinc cylinder Z stands, surrounded with water. A copper wire, covered with gutta-percha, proceeds from the copper plate K through the sulphate of copper solution to the next cell, where it is attached to a neck cast in the zinc plate similar to that shown in the figure.

These elements of Siemen's and Halske's vary in price from 2*s.* 6*d.* to 4*s.*, according to their size. Upon circuits offering great resistance they have been found to work very well indeed, and they are universally employed at the stations upon the Indo-European line between London and Teheran.

41. In America, Grove's battery was largely employed, and is still made use of to a considerable extent on the main wires between the leading offices; but it is now being rapidly pushed aside by a modification of Daniell's battery, known as the Callaud, introduced in France, and much used in that country. This is purely a gravity battery, having the solution of copper sulphate and the copper plate at the bottom of the vessel; over this, when the element is set up, is placed pure water, having the zinc plate suspended in it by means of small zinc hooks fixed to the sides of the vessel. This form, if frequently examined, works very well; but, in common with all its class, it requires very careful handling and treatment.

Other modifications of the Daniell used in America are those known as the Lockwood (in which the copper element consists of two flat spirals of wire united with each other) and the Baltimore, having a glass tube reaching nearly to the bottom of the jar, for the purpose of supplying copper sulphate when required.

Another battery lately introduced in America is named from its inventor the Eagle's Battery. The distinguishing feature in this is that the containing vessel is formed of lead, which is made to play the part of the negative element, zinc being as usual employed for the positive; the battery is said to give very fair results.

42. In India the Minotto, which is perhaps the best known and one of the earliest forms of gravity battery, is employed.

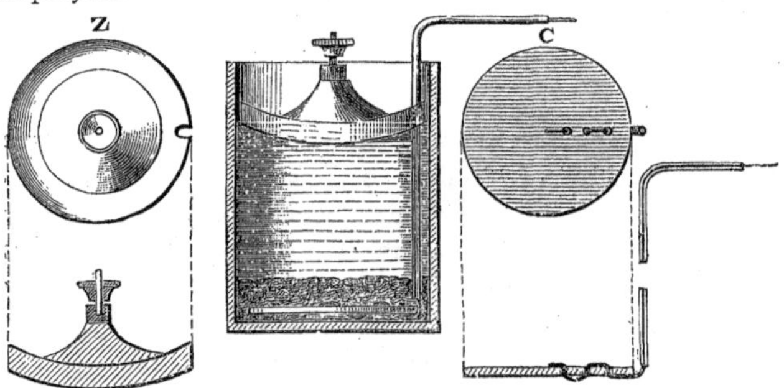

FIG. 15. $\frac{1}{6}$th real size.

It consists of a round earthenware glazed jar, in the bottom of which is placed a circular copper disc with three holes perforated in it, as shown in fig. 15.[1] Into these holes is slipped the conductor, of an insulated copper wire, which has been stripped of its covering for a distance of about $2\frac{1}{2}$ inches. This is then well hammered into the copper plate so as to ensure perfect metallic contact without the employment of solder; were solder made use of, local action would speedily ensue. Over the copper plate is packed from eight to twelve ounces of the copper sulphate, and above this is placed a piece of linen or blotting paper. Next comes a layer of moistened sawdust, or, in the event of sawdust

[1] The zinc plate is usually flat.

not being procurable, of clean river sand, which is to be preferred to the sea sand. This is likewise covered with a piece of blotting-paper, upon which finally rests the zinc plate, fitted with a brass contact piece, as shown in fig. 15. The insulated wire, whose conductor has been firmly welded into the copper plate, is led up through the copper sulphate and the layer of sawdust or sand, which is tightly pressed down, especially around it; it is then attached to the zinc of the succeeding cell. The whole is filled up with clean water to a height of about an eighth of an inch above the level of the zinc plate.

The connecting wire should be very carefully examined, and rejected if the trace of a flaw in the insulating covering is detected. No covering of tape should ever be employed, for the moisture spreading in time wholly over it, plays the part of a return wire, and places the cell on 'short circuit.' To prevent local action between the zinc plate and the brass contact piece, it has been found advisable to apply a coating of coal tar and resin to a point some little way above their junction.

This form of battery has been employed for many years in India, and has given every satisfaction; the number of cells in use at almost all the offices is comparatively so limited that they can all receive daily attention; and if the froth generated in the action be then drawn off and replaced by a little pure water, the cells will continue at work for a very long period. From eighteen to twenty months is the average life assigned to them; and it is stated that upon what are important lines they last sometimes for two years; and upon local lines, where little work is done, for as long as thirty to thirty-two months.

No other form of battery is used in India for speaking circuits. None but some other modification of the Daniell could successfully compete with it; for the means of transit are generally so slow and expensive that considerable inconvenience might arise were any materials employed in the battery which do not form a portion of the general stock-in-trade of the country. The copper sulphate possesses this advantage in being an article of commerce; it is manufactured amongst the natives, by whom it is largely employed for medicinal purposes, and it can be procured at very short notice whenever the necessity arises.

CHAPTER III.

SIGNALLING INSTRUMENTS.

43. TELEGRAPHY is the art of conveying to distant points the first elements of written language—either letters or numerals—by certain preconcerted signals; and the formation of these signals, by means of the action of currents of electricity upon permanent magnets, upon soft iron, and upon electrolytes,[1] forms the next portion of our subject.

44. Telegraphic signals are either *visible* or *audible*.

Visible signals, again, are either *permanent* or *transient*; in other words, they are either *recording* or *non-recording*; and they differ from each other either in form or position.

Audible signals, on the other hand, are always *transient* and *non-recording*; they differ from each other in tone or duration.

Hence we have different systems of telegraphy in which the signals are registered in different ways, and the currents do their work in different methods.

A. THE NEEDLE SYSTEM.

45. The needle is a visible system with transient or non-recording signals. It takes its name from the fact that the alphabet is formed by the vibration of a small pointer or

[1] An electrolyte is any compound substance which in solution is capable of being decomposed into its constituent elements by the passage of a current through it.

needle, movable between two fixed stops. N S, fig. 16, is such a needle, movable in the plane of the paper about its centre C, the distance of its motion being restricted by the stops *a* and *b*. This needle is capable of receiving two distinct motions, the one to *a* and back, and the other to *b* and back. Its normal position is vertical. In the earliest needle system five of these movable pointers were employed; the number was afterwards reduced to two, and this has been gradually superseded by, and now all but universally abandoned in favour of, a needle system where only one pointer is employed, and which therefore goes by the name of The Single Needle System.

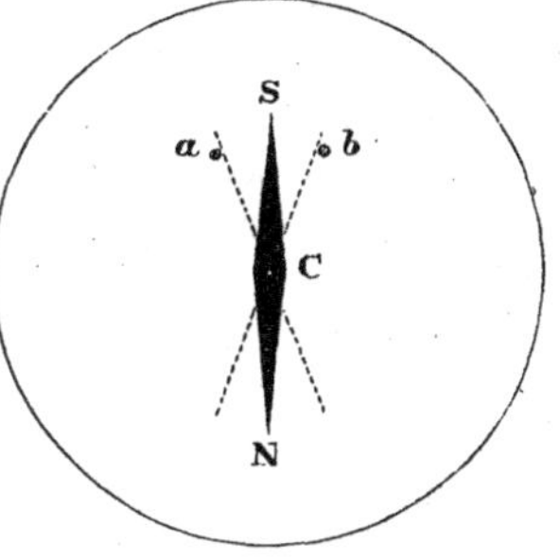

FIG. 16.

46. Since we have a motion to the right and a motion to the left as well, we can combine these two motions in any order or number we please, and so form a series of preconcerted signals which shall represent the alphabet. Thus, taking those letters which are most frequently used—viz. e and t—one motion to the left is the letter e, one motion to the right the letter t; and, taking those letters least used—viz. x and z—one motion to the right, two motions to the left, and one motion to the right, represent x; two motions to the right and two to the left represent

FIG. 17. ½ real size.

z. All the other letters are similarly formed of two, three, or four combinations; and thus, with a maximum combination of four movements of the needle, the whole of the alphabet can be formed. The manner in which the alphabet is made is shown by the above diagram (fig. 17), the little stroke \ representing a motion to the left and the longer stroke / a motion to the right.

47. The motion of the needle is produced by the mutual action of currents and magnets. Electricity and magnetism are so intimately related to each other that by many they are thought to be only different phases of the same agency. Thus the motion of a magnet always produces electricity; the transference of electricity always produces magnetism. The neighbourhood of a current is, in virtue of this fact, a *magnetic field*—a term introduced by Faraday to denote the entire space through which a magnet diffuses its influence—and a magnet or piece of soft iron placed there is influenced by the magnetism of that field.

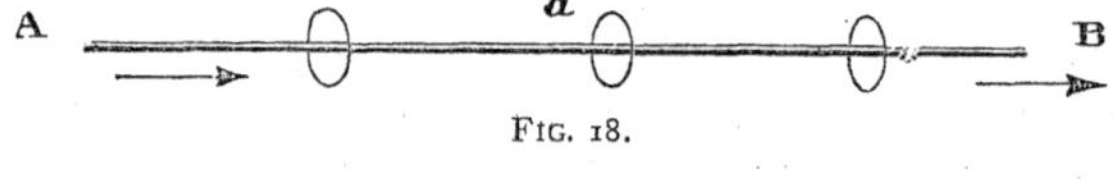

FIG. 18.

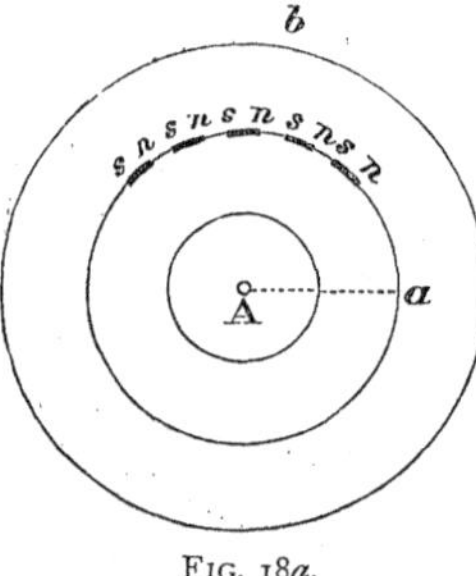

FIG. 18*a*.

Thus, if the wire A B, fig. 18, is traversed by a current in the direction shown by the arrow, it is surrounded along its whole length by a magnetic field; and if it be dipped while in this condition in a mass of iron filings, these filings will cluster around the wire, and adhere to each other in the manner shown in fig. 19.

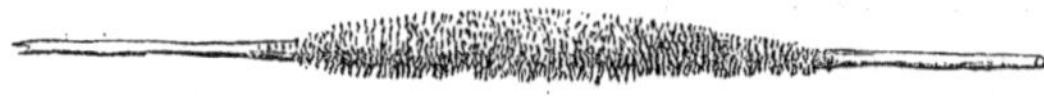
FIG. 19.

This is due to the fact that each little piece of iron acquires magnetism, and assumes the direction with respect to the wire which its *polarity* imparts to it. If a freely suspended magnet be placed in this field it will itself move in a certain direction, which is dependent on the polarity of the field. If in fig. 18*a* the wire A B, at any distance A *a*, were conceived to be surrounded by a ring of little magnets, all freely suspended by their centres, they would assume the positions shown in the figure, with all their N poles turned in one direction and all their S poles in the opposite direction. If their N poles were free, they would move in a circular path or orbit around the wire, in the direction shown. Hence we can conceive a wire conveying a current to be surrounded by a series of concentric tubes of magnetised matter, each formed by a series of concentric rings of magnetised molecules whose poles are all tangential, or at right angles to the wire. Such a series is shown in fig. 18*a*. Thus, if a magnet be brought within the neighbourhood of the wire it will be acted upon by the directive power of these imaginary magnetised molecules and tend to place itself at right angles to the wire, and always, under the same circumstances, in the direction shown by the little magnets in fig. 18*a*. We have conceived the current flowing in the direction A B. If it flow in the reverse direction, B A, the polarity of the field will be reversed (fig. 20). Hence a current in one direction will cause a magnet suspended above it to deflect to one side and in the opposite direction to the other side, and whenever the magnet is placed in the direction of the wire, it will always tend to form a tangent to a circle having that wire for a centre. There is no difficulty in remembering this direction of deflection. If you look at the face of a watch and conceive the current going *from* you, the N poles will all be 'negatively rotated,' or moved to the *right*, like the hands of the watch. The energy of this action between a current and a magnet depends upon the strength of the current passing, upon the strength of the poles of the magnet, upon its position and weight, and upon the distance between the magnet and the wire. The strength of the acting current upon the magnet can be practically multiplied at will.

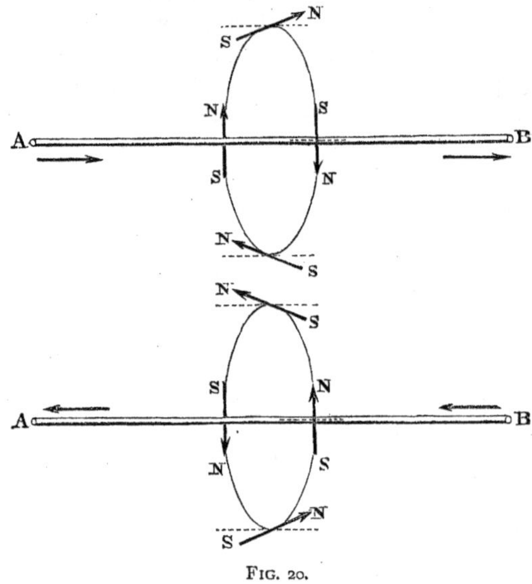

FIG. 20.

If the wire take a turn round the magnet, as shown in fig. 21, it will be evident, on a little consideration, that the directive action of the current as it passes below the magnet

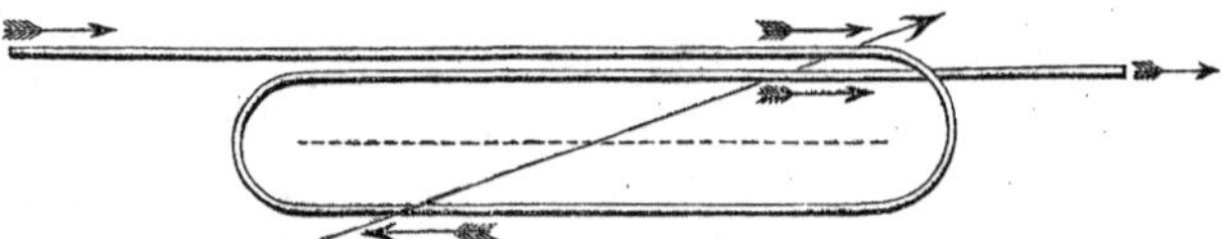

FIG. 21.

is the same as, and is added to, that of the current as it passes above the magnet: the effect of the current above the magnet is, in fact, duplicated by the additional turn. Hence the effect is triplicated in fig. 21. Thus, by multiplying the number of turns we multiply the effective action of the current upon the magnet. In this way we have the means of rendering sensible the presence of the weakest possible current, and we can, by varying the direction of the current, vary its

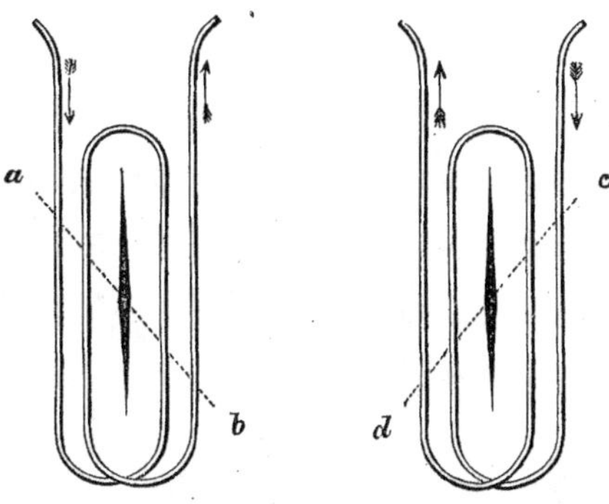

FIG. 22.

directive influence upon a magnet suspended along its length, so as to make it move either to *a b* or to *c d* (fig. 22).

48. The single needle instrument is based upon these fundamental facts.

There are two forms of the single needle instrument in general use, viz. the drop-handle and the pedal or tapper form. The essential principles of each are precisely the same; the only difference between them lies in the mechanism of the manipulator or sending portion of the instrument.

Fig. 23 gives a view of a tapper form of instrument, and 23*a* of a drop-handle single needle from which the external covering has been removed. A is the receiving portion of the apparatus. It consists of two ivory bobbins wound with fine silk-covered copper wire, and placed symmetrically with respect to a small magnetic needle free to move inside them. One end of the coils of wire is connected to line, the other to earth. Moving upon the same axis as this small magnetic needle is a steel indicator on the outside of the dial, fig. 17. The motion of this indicator is regulated by two small ivory stops placed at a distance of about half-an-inch apart upon a dial which is so constructed as to be capable of rotation upon its axis. If now a current of electricity pass through the

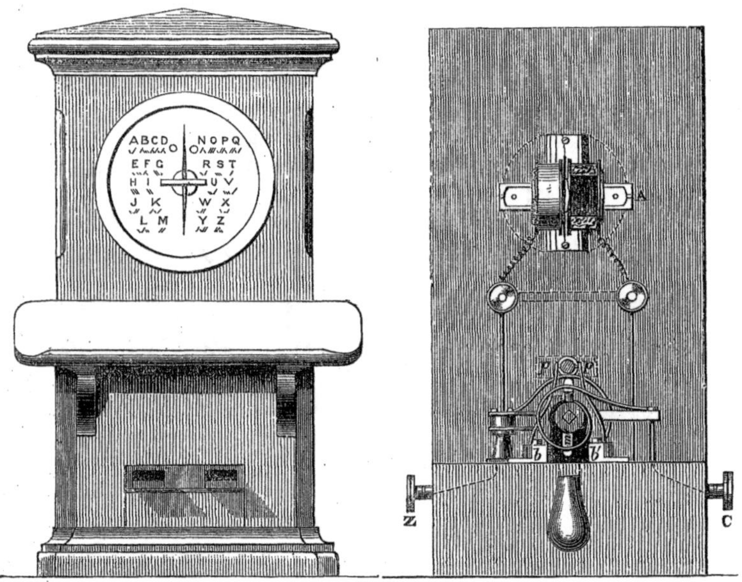

FIG. 23. $\frac{1}{6}$th real size. FIG. 23*a*.

coils of wire wound round the two small bobbins, the magnetic needle inside them will be deflected, and along with it the indicator outside. The direction of this deflection, whether to right or to left, will depend solely upon the direction in which the current is passing. The two coils are wound quite distinct from each other; but one end of the wire in each is soldered to the brass frame, and they thus act as if they formed one continuous coil. The advantage of this arrangement is that should the wire in either get broken or fused—as frequently happens by lightning—the circuit will still remain at work, provided the wire is carried over to either of the screws. All, therefore, that is necessary to enable communication to be kept up by this instrument, is an arrangement by means of which the magnetic needle can be deflected to right or left at will; in other words, an arrangement by which the direction of the current passing through the coils can be reversed when desired. An investigation of the mechanism of the commutator, or sending portion of the instrument, will show how this is carried out.

The wire from the copper pole of the battery is attached to the terminal marked C, that from the zinc to the terminal Z (fig. 24).

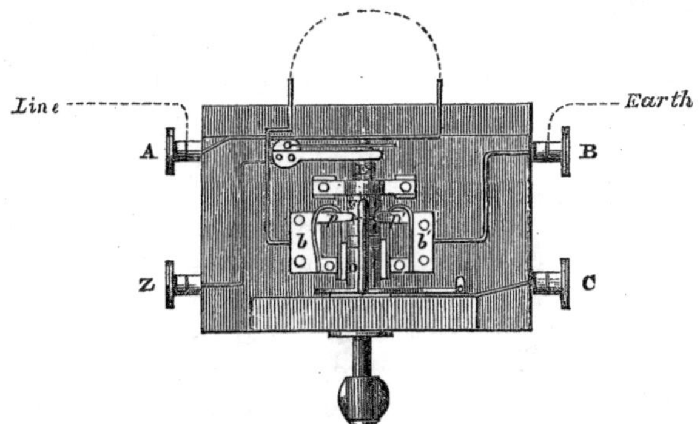

FIG. 24. ⅙th real size.

The line-wire is led to A, and a wire proceeds from B to the earth. The arbor of the handle D F consists of two parts, D and F, formed of gun-metal, and separated by some insulating material: ebonite, or more frequently box-wood, is employed. To D a wire leading from the copper terminal C is attached, to F a wire leading to the zinc terminal Z. *p*, *p'* are two steel springs, both of which are connected with separate brass bars, *b* and *b′*, in the base of the instrument; by means of one portion of brass-work *b*, *p* is in connection with terminal A through the coils, and *p′* is in connection with terminal B by means of the other portion of brass-work *b*. These two springs press against the 'bridge' shown at fig. 24, which maintains the continuity of the line. The half of the arbor F carries over it a metallic pin or projection *m*, which when the arbor is at rest remains between the two springs *p* and *p′* without touching either; whilst D is similarly fitted beneath with a pin or projection *m′*, which when the arbor is at rest remains between the two pieces of brass-work *b* and *b′*, and these are in connection with *p* and *p′* respectively.

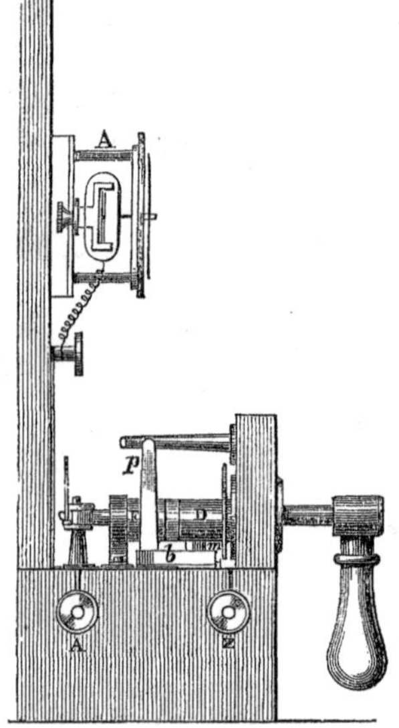

FIG. 25. $\frac{1}{6}$th real size.

Let now the handle be moved to the left: the projection *m′* of the half D moves to the left, and pressing against the brass-work *b*—which along with the spring *p* is in connection with A—brings the copper of the battery on to the line-wire; at the same moment the projection *m* of the half F is thrown to the right, and pressing against the spring *p′*—which with the brass-work *b′* is in connection with B—breaks its connection with the bridge and puts the zinc to earth. In this way a positive current is sent along the line, through the receiving apparatus at the distant station, deflecting the needle there, and returning by means of the earth to B and thence to the zinc of the battery.

Let the handle be next turned to the right. Everything

is reversed: the projection *m'* is now thrown into contact with *b'*, and thereby puts the copper to earth; *m* is meanwhile pressed against spring *p*, and thus brings the zinc to line. The current may now be regarded as passing along the earth through the coils at the receiving station, deflecting the needle in the opposite direction to what it previously did, and returning along the wire to A, whence it reaches the zinc of the battery.

The principle of the sending portion of the 'pedal' or 'tapper' form of single needle is as follows:—

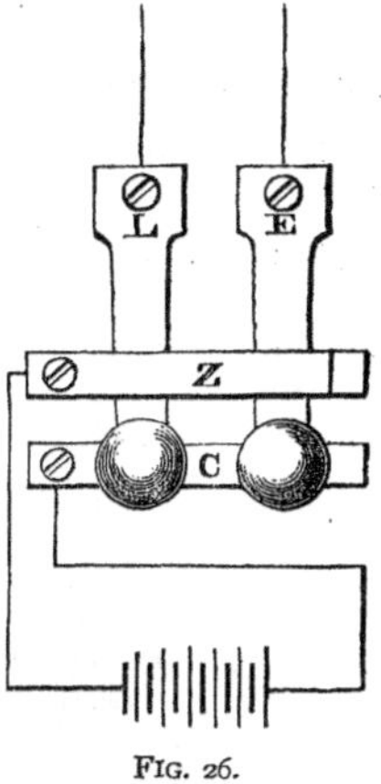

FIG. 26.

C and Z are two strips of metal to which the 'copper' and 'zinc' of the battery are respectively brought. E and L represent two metallic springs which are in connection with the 'earth' and line respectively, and which, when at rest, press against Z. If now L is depressed and brought into contact with C the circuit is completed, and the current starting from C traverses the line wire and the coils of the receiving instrument at the distant station, returning by means of the earth to E, which, being in contact with Z, is in connection with the zinc of the battery. If, on the other hand, E is depressed after L resumes its normal position the direction of the current is reversed, for the copper of the battery is now to earth and the zinc to line; consequently the needle at the distant station is deflected in the opposite direction.

49. The details of the construction of the pedal arrangement cannot be faithfully represented by a diagram, but no difficulty will be found in comprehending them if the principle stated above is clearly carried in the mind.

50. The single needle is essentially an English instrument; it was invented and is still largely employed in

England, especially upon the railways, where no other form of instrument has ever been able to compete with it. The adjustment of the receiving portion of the apparatus is of the simplest possible character; in fact, when once at work no adjustment whatever is required. Any reasonable number can be joined up in circuit upon the same wire without fear of a complaint as to their working, unless it may be that of weak signals; and this can be readily obviated by the employment of additional battery power. The main defect in the older form of instrument was the liability of the small magnet inside the receiving coils to be partially, sometimes entirely, demagnetised, and even reversed in polarity by lightning. Mr.

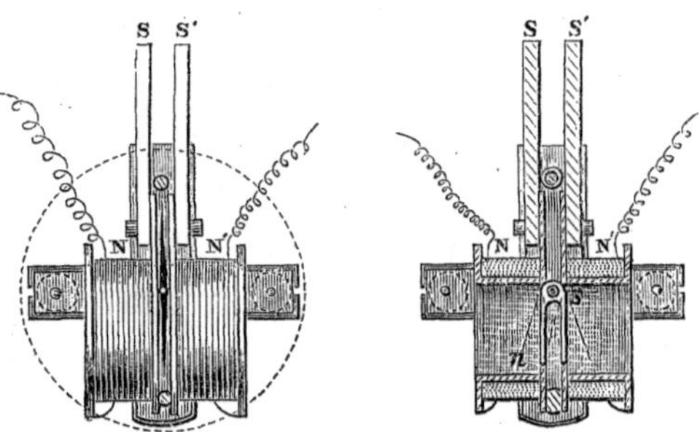

FIG. 27.

S. A. Varley has however entirely got rid of this danger, by the introduction of what is unquestionably the greatest improvement recently effected in this class of instrument. Instead of a small permanent steel magnet a soft iron needle, of the shape shown in fig. 27, N S, is employed. This owes its magnetism to the influence of two permanent bar magnets, N S and N′ S′, whose like poles are adjacent to each other, and which are fixed each into a brass frame around which the wire is wound. These bar magnets are very seldom demagnetised by lightning, except during storms of exceptional violence. They, however, lose their magnetism, like all permanent magnets, after the lapse of time, and ought to be remagnetised occasionally.

51. The principle of the double needle system is identical with that of the single needle. Two wires are employed, and every part of the apparatus is duplicated. The alphabet requires fewer combinations, and the instrument is, therefore, quicker in its action.

It is only necessary to illustrate this system by the subjoined diagram :—

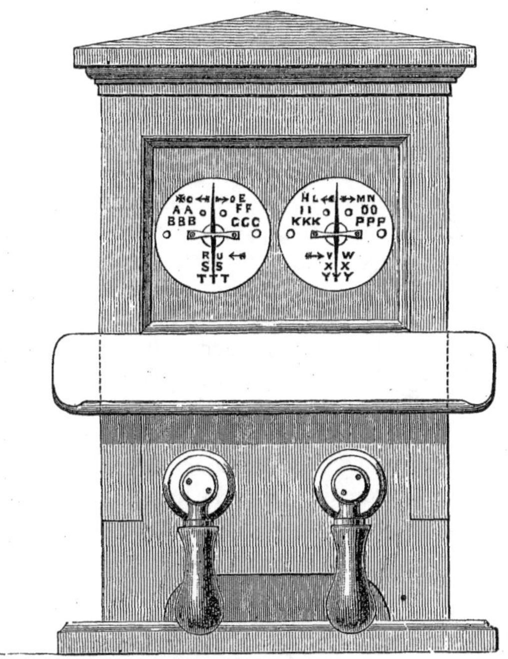

Fig. 28. ⅛th real size.

Many hundreds of these instruments still remain in use upon the English railways, but they are rapidly being supplanted by the single needle, which is much more economical and convenient in its arrangements.

B. THE ACOUSTIC SYSTEM.

52. The acoustic system is, like the needle, a transient or non-recording system, but differs from it, as its name implies, in the fact that the ear is made use of instead of the eye to interpret the signals sent. There are two forms of instrument employed in working this system, viz., *the Sounder* and *the Bell.*

53. Both these instruments are based upon the electro-magnetic effects of the current. Inasmuch as the neighbourhood of a current is a magnetic field, and filings of iron placed in that field acquire magnetic properties (§ 47), it follows that if we envelope a mass of iron filings—or even better, a piece of iron itself—with a ring of wire conveying a current, every filing or molecule of iron within this circle will be similarly magnetized; that is, it will be so magnetized that similar poles lie in similar directions. Let A B (fig. 29) be such a wire, conveying a current in the direction shown, and S a flat disc of soft iron. Now inasmuch as every molecule constituting the soft iron disc lies in the magnetic field of that current, it will be polarized in the direction shown at *b* in fig. 30; and as all these molecules have their polarities in

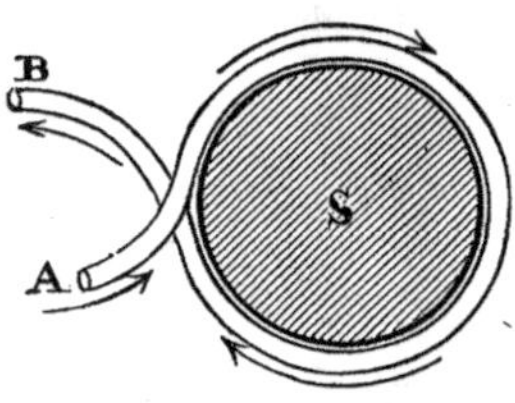

FIG. 29.

FIG. 30. FIG. 30*a*.

the same direction, the resultant effect is as though there were one magnet whose N pole was above and S pole was below the disc. Moreover, if instead of one ring of wire we were to surround the disc with several rings, the current flowing in the same direction in each ring, as shown by fig. 30*a*, the effect would be magnified; and if we were to superpose

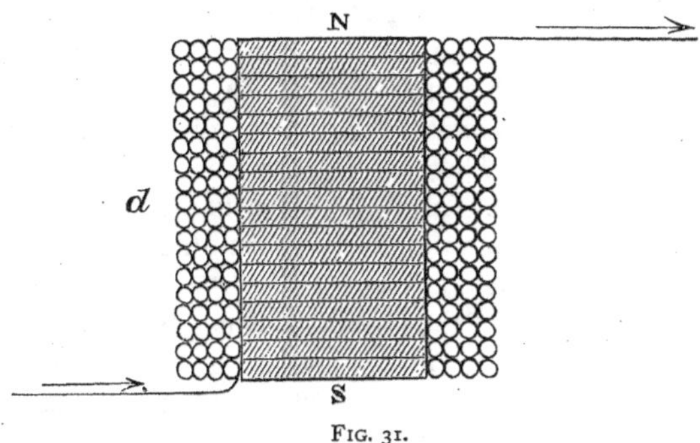

FIG. 31.

several discs, as in fig. 31, thus surrounded with rings, in all of which the current flowed in the same direction, the effect

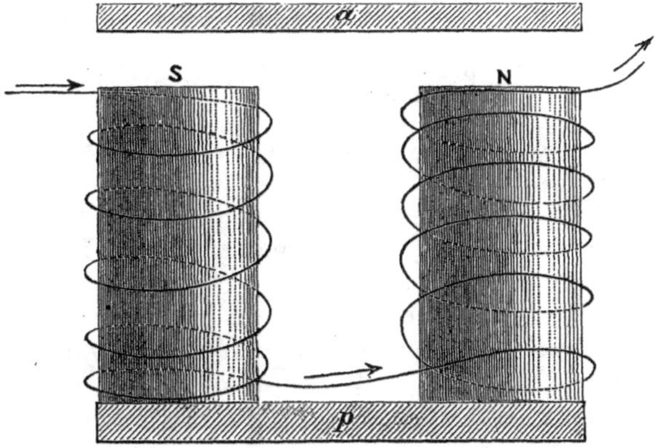

FIG. 32.

would be still further magnified, and we should have a powerful bar magnet, N S. Precisely this effect is produced by winding a helix of wire around an iron bar or core, and by combining two such iron bars (fig. 32) by a cross piece of soft iron *p*, and surrounding each bar with a coil of silk-covered wire, we construct an electromagnet which is powerfully magnetized every time a current flows, and which therefore exerts attraction upon a bar of soft iron or armature *a* placed in front of it. The power which this electromagnet exerts depends upon the strength of the current flowing, upon the number of turns the wire takes around the core, and upon the size of this core. Thus a very powerful current requires but a few turns of thick wire to produce its maximum effect, while very weak currents require a great number of turns of very fine wire to produce any effect at all.

54. The direction of the poles of the magnet is dependent upon the direction of the current and upon the direction in which the helix is wound. Electromagnets are almost invariably wound with the right-handed helix, shown symbolically by fig. 33, and the polarity due to the different directions of the current is shown by *a* and *b*. Thus, if the current flows around the iron core in the direction of the hands of a watch whose face is held before the eyes, the N pole is away from us.

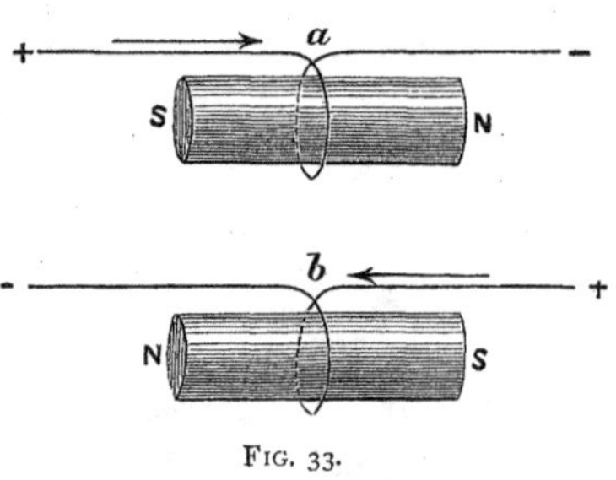

FIG. 33.

55. We can make the armature the end of a lever *a b* (fig. 34), pivoted at C, and we can limit its play by the two adjustable screws *d* and *e*. We can also maintain the lever in its normal position pressing against *e* by means of the antagonistic spring S, so that whenever a current passes through the coil, whatever its direction may be, the attraction of the magnetized core overcomes the tension of the spring,

and causes the end of the lever *b* to make a sharp blow against the adjusting piece *d*, and take up the position

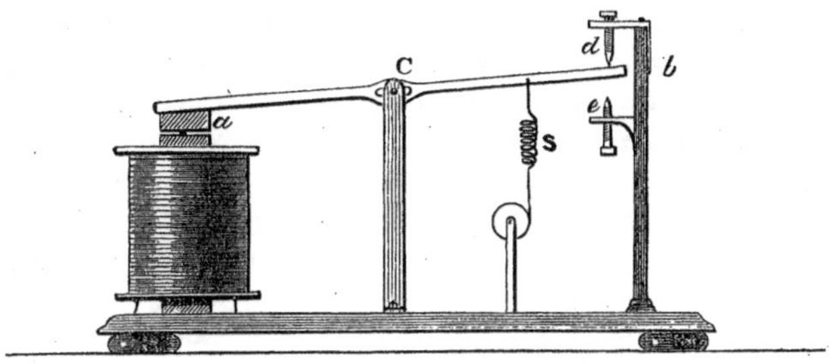

Fig. 34.

shown in the figure. When the current ceases, the attraction also instantly ceases, and the lever is pulled smartly back into its normal position against *e* by the action of the spring s. The blows made by the lever against *d* and *e* emit distinct and clear sounds, which are taken advantage of to convey to the ear the letters of the alphabet and other preconcerted signals. This is the principle of the Sounder.

1. *The Sounder.*

56. How can we convert the sounds made by the contact of the lever against the two limiting stops into an alphabet. We have shown (§ 46) how two motions, a motion to the right and a motion to the left, of a vertical needle have been applied to the communication of preconcerted signals through the eye. If we make one kind of sound to represent the motion to the left and another kind of sound to represent the motion to the right, we can do the same thing through the ear. In the sounder the sounds themselves are alike, but we can make one signal sharp, quick, and clear like the *quaver* note in music, and the other signal longer, slower and more deliberate, like the *crotchet* in music. The two sounds made by the lever in

striking the stop pieces *d* and *e* (fig. 34), enable us to do this. The lever striking *d* gives the commencement of the signal, and striking *e* the end of the signal. The time elapsing between these two sounds determines the kind of signal. Representing the one signal by a dot (.), and the other by a dash (—), we have the dot and dash alphabet of Morse.

57. It will be seen that in this alphabet we have introduced *duration* as an element of signalling. There is a very important element, viz. duration of silence as well as duration of sound. The alphabet is formed of short and long audible signals, *dots* and *dashes* separated by variable intervals of silence or *spaces* as they are called. There are three kinds of spaces : the space separating the elements of a letter, that separating the letters of a word, and that separating the words themselves. These durations of silence are as necessary as the durations of sound to perfect telegraphing on this system. Thus sound reading and sending is a method by which time is divided into accurate multiples of some arbitrary standard or unit—viz. the dot.

1. A dash is equal to *three dots.*
2. The space between the elements of a letter is equal to *one dot.*
3. The space between the letters of a word is equal to *three dots.*
4. The space between two words is equal to *six dots.*

The basis of the alphabet therefore is the dot

. representing the letter .... e

and the dash

— representing the letter .... t

Placing a dot before each of these elementary characters, we have

. . .. i

. — .... a

Placing a dash before each elementary signal, we have

— · .... n
— — .... m

Now affixing to each of the above four signals first a dot and then a dash, we have

· · · .... s
· · — .... u
· — · .... r
· — — .... w
— · · .... d
— · — .... k
— — · .... g
— — — .... o

Pursuing the same system with these eight characters, we have

· · · · .... h
· · · — v
· · — · .... f
· · — — ....(German) ü
· — · · l
· — · — ....(German) ä
· — — · .... p
· — — — .... j
— · · · .... b
— · · — x
— · — · c
— · — — .... y
— — · · .... z
— — · — .... q
— — — · .... ö
— — — — .... ch

There is also the French accented è · · — · ·, but with this exception no letter exceeds four signals.

A combination of five signals is employed to represent the numerals and cypher

1	· — — — —
2	· · — — —
3	· · · — —
4	· · · · —
5	· · · · ·
6	— · · · ·
7	— — · · ·
8	— — — · ·
9	— — — — ·
0	— — — — —

The stops and other signs of punctuation are made by a combination of six signals.

Period or full stop	· · · · · ·
Repetition or ?	· · — — · ·
Stroke, or the divisional bar of a fraction	— — — — — —
Hyphen	— · · · · —
Apostrophe	· — — — — ·

There are many other signals in use, but they are exceptional. Some of those indicated above are rarely employed in England. *Ch*, for instance, has been abandoned because it is so much like '*to*.'

58. Fig. 35 represents a simple sounder arranged for the conveyance of the above signals to the receiver, and is the one which is at present generally employed in England. It is in every respect the simplest of all the signalling apparatus in use, and simplicity in construction is a great consideration when technically unskilled operators are employed. One end of the wire of the electromagnet is attached to a brass terminal fixed in the wooden frame : the other end is attached to a similar terminal. To these same terminals the

line and earth wires are respectively brought. S is the antagonistic spiral spring, having one end attached to the lever

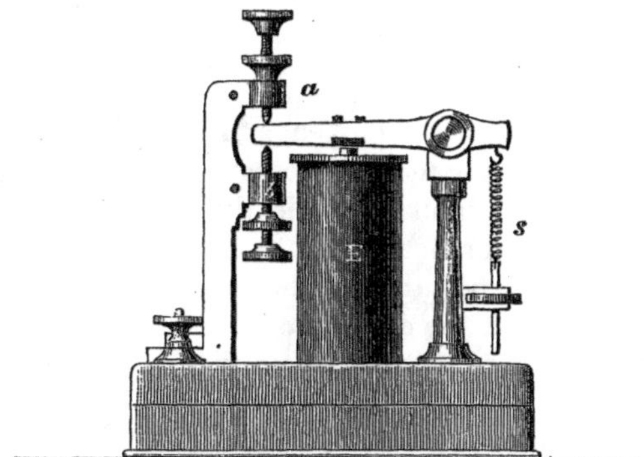

FIG. 35. ¼th real size.

and the other to an adjusting screw, as shown in the figure, by means of which its tension may be increased or decreased at pleasure, so as to compensate for the variation in the strength of the line current. *a* and *b* are the two adjustable stops against which the lever strikes, whence the sounds that are read are emitted.

59. *The Key.*—How are these currents of various duration sent by the sending station? The apparatus for doing so, called *a key*, is much simpler than that required in the needle system, because no reversals are needed, currents in only one direction being required. The key (fig. 36) consists of a simple brass lever K, which is in connection with the line wire, and which is pivoted so as to be movable about its centre on a brass piece fixed upon an insulating board of wood or ebonite. It is maintained in its normal position by a spiral or bent spring not shown in the figure, causing the back of the key to be held in contact with the back stop 3, which is in connection with earth, thus preserving the continuity of the line wire and earth. One pole of the battery is placed in

connection with the front contact piece 1, the other being put to earth. Thus, whenever the key is depressed, 2 is brought into contact with 1, and the current flows to line. The moment the key is raised the contact is broken, and the current at the same moment ceases. The duration of the current thus evidently depends upon the duration of this contact. These currents pass through the receiving instrument at the distant station and operate the sounder in the manner described in § 56. Hence to send dots and dashes by this key it is only necessary to tap or move it as one would the key of a piano in order to produce crotchets and quavers.

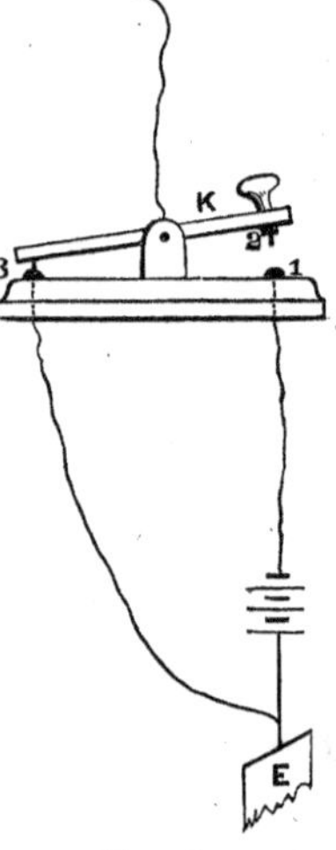

Fig. 36.

60. Such is the sounder in its simplest form; and though it is not always possible to work it in this simple form except for very short distances, yet such is the principle on which it is constructed and worked. When we have described other systems, we will draw a comparison between the advantages and disadvantages of the different plans in use, and show why it is that the sounder is gradually supplanting other forms of signalling.

61. *The Relay.*—But as we have said, it is not always possible to work it in this simple form. As the lines increase in length, and the effects of imperfect insulation make themselves felt, the line currents become so weak that they are unable to produce audible signals upon our sounders, or batteries have to be employed so powerful that they do more mischief than good. Some method is then needed to bring in fresh currents which will make the signals audible. This is the function of *the Relay*, by which a local battery is brought

into play which works the receiving sounder in the same way as the line current would have done had it been of the requisite strength.

62. The relay is, in fact, nothing more or less than a more delicate form of the electromagnet employed in the sounder previously described. It is wound with a finer and longer wire (§ 53); all its parts are more delicately constructed, so that it will move with the weakest currents. However long a line may be, and however badly it may be insulated, as long as any currents can get through, so long can relays be constructed to move with those currents.

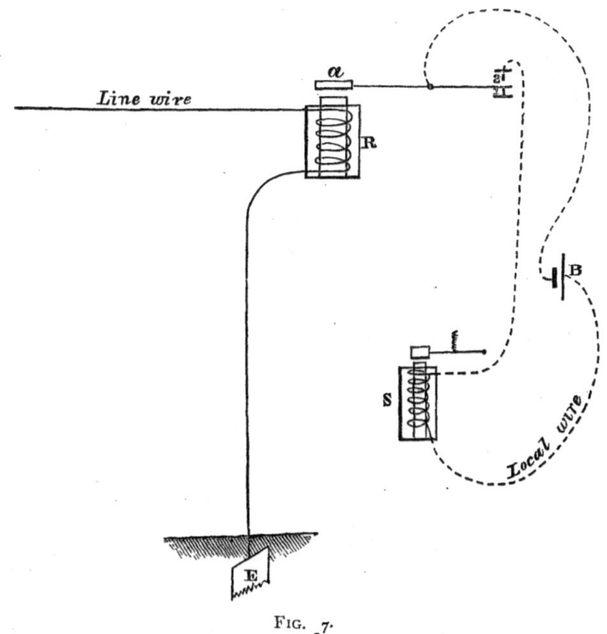

Fig. 37.

The principle of its operation is given by fig. 37. S is the electromagnet of the ordinary sounder wound with thick wire and giving but little resistance, and which therefore cannot be worked by the line current. R is the electromagnet of the relay wound with very fine wire, and worked by the line current. B is a local battery whose copper pole is attached to one end of the wire of the sounder, and whose zinc pole is connected with the lever of the relay. The other end of the coil of the sounder is connected with the upper contact 2 of the relay. When a line current passes through R it attracts the armature *a*, brings the lever in contact with 2, and completes the *local circuit.* The local current therefore works the sounder, whose armature remains in contact just so long as that of the relay does, and thus every movement on the relay is repeated on the sounder.

63. There are many different forms of relay. Such a one as that just indicated is called a *non-polarized* relay, but it is not much used in England. The forms of relay more largely used are called *polarized*, because their armatures are either permanent magnets or are maintained in a magnetized condition by permanent magnets. They differ principally from the non-polarized relay because they are affected by the direction of the current, and under certain circumstances they are far more sensitive.

64. *Siemens' Relay.*—A sectional view of this apparatus is shown by fig. 38, and a plan of the top by fig. 39.

N S (fig. 38) is a hard steel permanent magnet, into a slit in the upper or S end of which is pivoted a soft iron armature *a b*, capable of motion in a horizontal plane about the centre *b*, and having a small aluminium tongue C fixed on to its free end; on the lower or N end of N S rests an iron bar A, which supports the two soft iron cores of the electromagnet E; the further extremities of these are terminated in the pole pieces P and P_1 (fig. 39), which are fixed by screws and can be moved to and fro at will. D and D′ are two contact points whose position can be varied by means of the adjusting screw B; when the tongue C presses against the former, the local circuit is completed; when drawn against

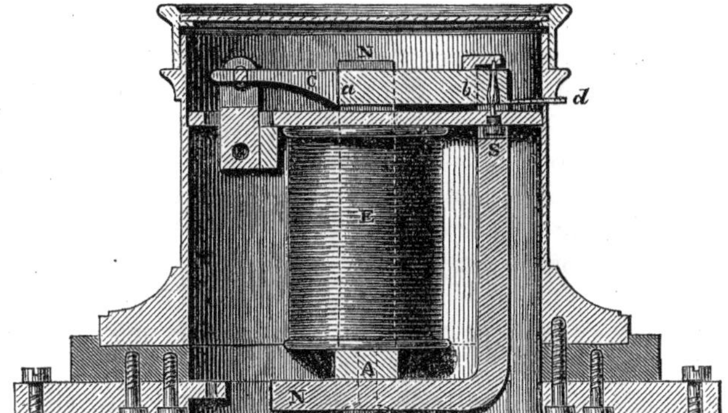

FIG. 38.

FIG. 39. ½ real size.

the latter, which terminates in an insulating agate point, it is broken. The coil wires of the electromagnet are attached to terminals 5 and 6 ; to that marked A a wire from the zinc of the local battery is brought—the copper being carried to one end of the coils of the instrument which it is intended to work, whilst to the terminal, here marked B, the further end of the instrument coils is connected.

The action of the relay is as follows :—

The end N of the permanent magnet N S induces S polarity in the bar A and the ends of the cores next to it, but N polarity in the upper ends remote from it and terminating in $P P_1$, both of which are therefore N poles. The end S, on the other hand, induces N polarity in that portion of the armature *a b* next to it, and S polarity in the further extremity, moving between P and P_1 (C being a non-magnetic metal is not affected). When, therefore, A B is equidistant from P and P_1, it is equally attracted by both, and may be supposed to touch neither D nor D′. If the pole P be approached nearer to *a b*, it obeys its influence and is attracted to the agate point in D′. This is the position of the armature when the relay is at rest.

As soon, however, as the line current enters the coils from terminal 5, $P P_1$ becomes an electromagnet whose N pole is P and S pole P_1 ; the pre-existing north polar magnetism of P is consequently very much increased; that of P_1 is on the other hand greatly diminished, and, if the line current has been of sufficient strength, is entirely neutralized or even reversed. The result is that under the influence of a more powerfully attracting force, C is drawn from D′ to D, and remains there so long as the line current is flowing, returning to D′ when this ceases. In this way the local circuit is completed, and the instrument worked in exactly the same manner as though a line current of equal strength to that of the local current had been the cause. D is connected by means of a wire passing through the inside of the relay to B, whilst A is attached to the brass work of the frame, and by means of it is placed in direct communication with the armature *a b* and tongue C.

The adjustment of the apparatus is extremely simple, and consists at the outset of three distinct movements.

1st. P and P_1 are so placed that the armature *a b* shall remain as nearly as possible exactly between them.

2nd. D is next turned so that the armature shall have a play of about $\frac{1}{25}$ of an inch between D and D′.

3rd. B is then turned so that the armature shall press gently against the insulating stud D′.

The screw B is the main regulator when the apparatus is at work: if from the line current being very strong the dashes are continuous or the dots too long, turn B to the right, thereby bringing *a b* a little harder on to D′; if this has not the desired effect, lessen the power of P by moving it a little further back or by throwing P_1 forward. If, on the other hand, owing to the weakness of the line current, dots are lost, turn B to the left, thus bringing *a b* nearer to D; and if this does not prove sufficient, bring P a little forward and throw P_1 further back.

In most instances it will be found that B does everything which is required. The simplicity of adjustment, the little likelihood of the apparatus being demagnetised on account of the large inducing magnet which is employed, except after the lapse of several years, and the general compactness of this instrument, recommend it for general use. The only objection that has ever been urged against it, is that it is scarcely sensitive or light enough for very long lines, rapid sending, and extremely weak currents, on account of the comparative weight and inertia of the mechanism. A relay manufactured by Mr. Stroh leaves little to be desired in point of delicacy, and is well adapted for long circuits, although for general working nothing can surpass the form which has been described.

65. *Stroh's Relay.*—A plan of Stroh's relay is shown by fig. 40, and a diagram of the interior by fig. 41.

The essential difference between this form of relay and

that manufactured by Messrs. Siemens is that the large permanent inducing magnet of the latter is dispensed with, and in place of the induced armature a light permanent compound magnet is employed. This is shown in fig. 41.

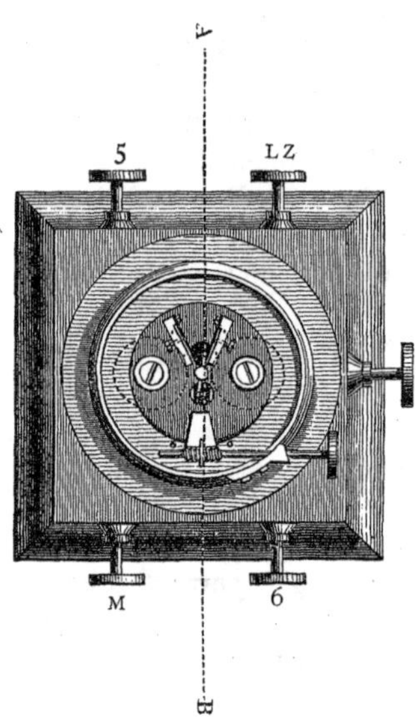

FIG. 40. ¼th real size.

Two small permanent steel magnets, N S and N′ S′, of the form here given, are placed with their opposite poles adjacent to each other; they are fixed in metallic connection to a steel axis A B, and are movable along with it.

The electromagnetic arrangement likewise differs from that in the Siemens' relay; the two cores instead of being united by a piece of soft iron and thus forming one electromagnet, are joined by a small bar of brass, and become when magnetised by the current two distinct bar magnets, whose unlike poles, owing to the wire being wound in different directions, are opposed to each other.

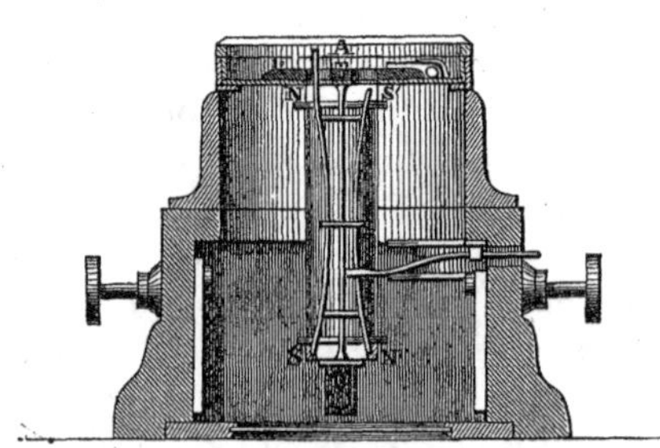

FIG. 41. ¾th real size.

The lower end of the axis of the compound magnet is pivoted into the brass bar; the upper end into the brass top of the apparatus. The two contact points between which the upper extremity of the longer magnet plays are shown in fig. 42; they are each fitted with a steel spring s and s′, whose position is regulated by two small screws. The electrical connections are the same almost as in the Siemens' relay; to terminal 5 (fig. 40) one end of the coil wire is attached, the other to 6. The local zinc is led to L Z, and from there it goes by means of a wire inside the covering to the brass work of the instrument, with which the compound magnet, as already observed, is also in connection; to M comes the wire from the instrument which the local current is intended to work, and this is continued to s′ (fig. 42). When the apparatus is at rest, and no current is circulating through the coils of the electromagnets, the end N of the compound magnet is adjusted to press against s, but immediately the line current is received, then the two cores become two powerful magnets; the mutual attraction and repulsion between them and the two small magnets forming the compound magnet cause the end N of the latter to be drawn from s to s′, and take up the position shown in fig. 42. N being in connection with the local zinc and s′ with M, the local circuit is thus completed. Immediately the line current ceases to flow, N falls back to its normal position against s.

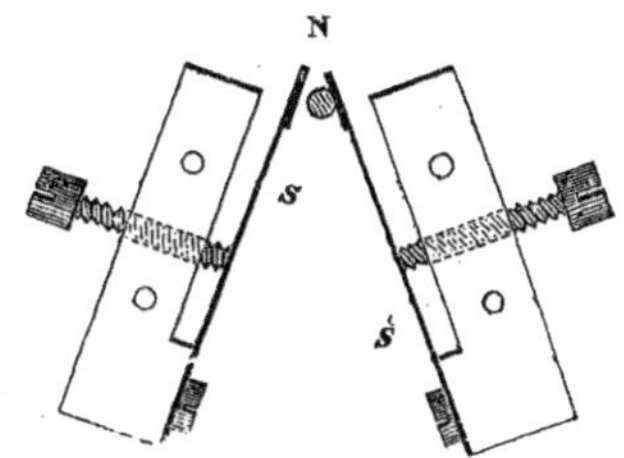

FIG. 42. Full size.

The double action of the electromagnets upon the compound magnet, combined with the comparative lightness and ease of motion of the latter, enables this relay to be worked with very delicate currents, especially when the double current system (§ 108) is employed; the small play of the magnet between the contact points is likewise of advantage where fast speed instruments are employed, and in these cases it is now coming to be generally adopted in England.

In one important respect this Stroh's relay is decidedly inferior to the Siemens':—the light compound magnet has the advantage of quickness of motion, but it is obtained at the expense of safety in working and of permanency of magnetism. Lightning frequently demagnetises, or even reverses, the compound magnet in Stroh's relay: no case has ever occurred, in England at least, of this happening to the magnet in Siemens' relay. Also the little magnets soon lose their magnetism if worked in connection with either of the systems described.

2. *The Bell.*

66. With either a Siemens' or a Stroh's relay sounders can be worked at any distance, and in England through any weather. The sounder was introduced in America, and it has there supplanted all other forms of apparatus. It is also almost universally employed in India. But the earliest form of acoustic instrument introduced into England was Bright's Bell. Two bells of different tone are used, the hammer of one being actuated by currents in one direction, and that of the other by currents in the other direction. The sound of one bell corresponds to dots, and that of the other bell to dashes. The sending apparatus is the same as in the pedal

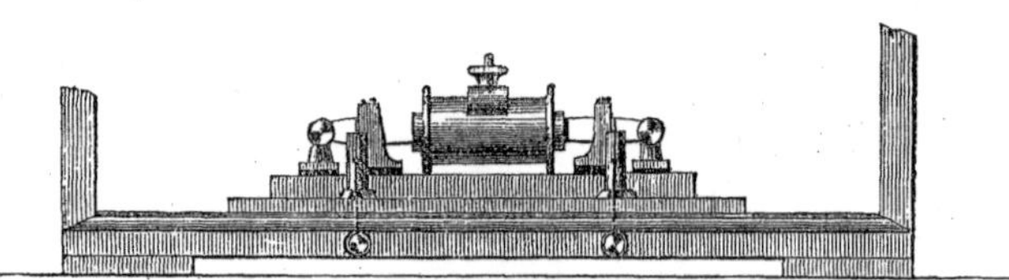

Fig. 43. $\frac{1}{5}$th real size.

single needle, and relays and local currents are needed. The instrument is very quick in its action, probably the quickest non-recording instrument extant, but it is complicated in its construction and difficult in its adjustment compared with the sounders. It is rapidly being supplanted by its fitter opponents. Its general construction is shown by figs. 43 and 44, the former being a back view of the relay.

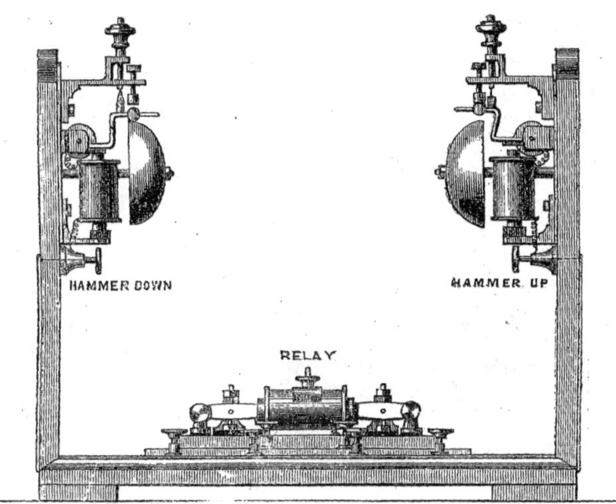

FIG. 44. ⅛th real size.

67. In the instruments which have been described the signals have been transient, and have left behind them no permanent record for reference. We have now to deal with recording instruments in which the signals are permanent.

The simplest and earliest of all is the 'Morse' recording instrument, so called from its inventor.

C. MORSE SYSTEM.

68. *The Embosser.*—The first form of Morse recorder was the Embosser, shown by fig. 45.

The radical principle is exactly the same as that of the

sounder, which has been already described. The recording arrangement is purely mechanical, and is as follows:—

E E′ are the coils of the electromagnet as before, O the armature. To the latter is attached a lever movable around an axis, and carrying at its further end a small steel style S. When the armature is attracted, and with it the end of the lever drawn down, this style is thrown upwards and pressed against a strip of paper *p*. This strip of paper is made to unwind itself by being passed between two friction rollers W W′, which are set in motion by the action of clockwork. In the upper roller and just over the style is a small groove, into which the paper is pressed so long as the armature is attracted. A mark is thus embossed on the upper surface of the paper, which will appear in the form of either a dot or a dash according to the time that the armature has been held down, and the style elevated; these, it will be seen, correspond, to the short or long sounds in the simple sounder.

FIG. 45. ⅙th real size.

69. The reading of the signals made by the embosser is so fatiguing to the eye, that the instrument has now been entirely supplanted by the more modern form of recorder, viz. the Ink-writer. The first instrument of this description

was invented by Thomas John, an Austrian engineer, in 1854. The main object which he had in view was to reduce as far as possible the force which was required to drive the style on to the paper before the marks could be distinctly recorded in the embosser. He succeeded in doing this by substituting in place of the style a small metallic disc, which was kept constantly revolving in the inking fluid, and which was pressed against the paper as it unwound itself, according to the time that the armature of the electromagnet was held down. All the ink-writers which have been brought out since 1854 have been simple modifications of this idea, and the most perfect instrument which is now in use is only a mechanical improvement upon John's original principle.

Various arrangements have been tried for the purpose of increasing as far as possible the delicacy of the apparatus. The best is that which was introduced by Messrs. Siemens and Halske, and is now almost universally employed wherever recording instruments of this class are in use.

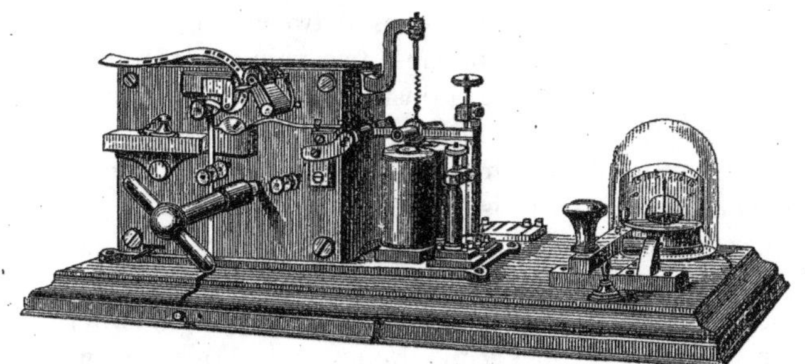

FIG. 46. ⅛th real size.

Fig. 46 shows one of the latest forms of these, and fig. 47 gives a section of the electrical portion of the receiving apparatus.

70. *The Ink-writer.*—E (fig. 47) is the electromagnet, which is worked in the same way as the sounder, already described; F is the armature; S is the antagonistic spring,

whose tension may be increased or diminished at will by means of the screw C. To F is attached the lever *f*, movable upon the axis at G and carrying the small disc I at one

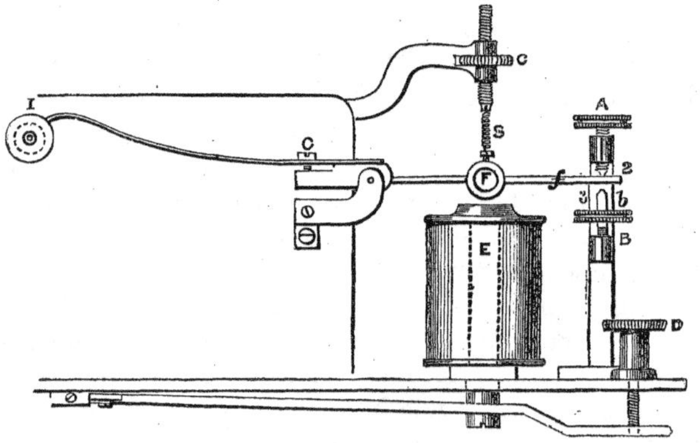

FIG. 47. ¼th real size.

end, whilst the other end moves between the two points marked 2 and 3: either of these two points may be raised or lowered at will by the adjusting screws A and B: the disc I dips into a reservoir of ink. The paper is wound upon a roller fixed in a drawer in the base-board of the instrument, and its path is indicated in fig. 46. It passes between two friction rollers, which are set in motion by means of an ordinary clockwork arrangement, which is enclosed in the case, and is liberated by the movement of a switch.

In addition to moving the friction rollers this clockwork arrangement also causes the disc I to revolve in the opposite direction to which the paper runs, in the ink-well, and in this way I is kept constantly wet with ink so long as it is required. When F therefore is kept down for a short space of time, I is thrown momentarily up against the paper strip and records a dot upon it: a dash in like manner is recorded if F is kept down for a longer time.

The paper which is employed is slightly coloured, and is cut into strips of about half an inch in width: the ink is ordinary printer's ink of good quality, diluted with olive-oil from which the stearine has been removed by freezing.

71. This instrument has four distinct and separate adjustments :—

1. Screws A and B, which regulate the play up and down of the armature F, and therefore of the inking disc I.

2. Screw C, which regulates the tension of the antagonistic spring S, tightening or slackening it as may be required.

3. Screw D, which regulates the distance between the poles of the electromagnet E and the armature F, by raising and lowering the coils, so as to increase or diminish at will the attractive force between the two.

4. Screw G, which regulates the position of the inking disc, with respect to the paper and armature.

It is regulated for working thus :—

I. (*a*) The screw B is first adjusted, so that the disc I gently touches the paper without pressing it too hard when the brass stop *f* is placed in contact with the stud *b*.

If the disc presses the paper too hard, it makes thick and indistinct signals : if it presses too light it causes the disc to jump and signals to split : thus — may become · — or – —.

(*b*) The electro-magnet is then raised by turning the screw D to the right, so that when the brass stop *f* rests upon the raised stud *b*, the poles *just* clear the armature *without actually touching it.*

A thin streak of light should be seen between the armature and the poles of the electro-magnet.

(*c*) The screw A is next adjusted so that the brass lever *f* is allowed to move through a space of about $\frac{1}{16}$ of an inch. A and B together so regulate the play of the inking disc that while it just dips into the ink-well it also *gently* presses against the paper, so as to mark it clearly.

II. The screw C is now adjusted so that the tension of the spiral spring S is just sufficient to bring back the armature to its upper position of rest when the current ceases to act, and to overcome the effects of *residual magnetism*, if any exist. The amount of tension required varies with each instrument. The letter V generally should be sent by the distant station, using very low power, and the spring gently tightened until the marks are clear and distinct. The ear is usually the best judge of this adjustment.

The instrument is thus regulated to work with the most delicate currents.

III. The coils are lowered by means of the screw D until a space of $\frac{1}{8}$ of an inch separates the armature from the electromagnet, and the nearest station should send V's with full power, and the coils should be raised until marks are distinct and clear.

The instrument is thus placed in a position to work with every current sent, viz. with the strongest current from the nearest station, and with the weakest current from the most distant station, provided there are more than two stations upon the same wire.

IV. The intermediate adjustments are made by the screw D, and as the currents vary so must the distance between the armature and the electromagnet be varied ; hence the only adjustment necessary to meet currents of different strength from different stations, or from the same station, during different states of the atmosphere, when once the instrument has been placed in working order, is that effected by the screw D.

(*a*) If, when a station is working, a continuous mark is made upon the paper, or signals run into each other, thus— ——— (usually the result of a powerful current), the coils should be lowered by means of D until marks are clear and distinct.

If marks should still run together when the coils are

as low as they can be, then the antagonistic spring must be tightened up.

(*b*) If marks fail (dots are lost, and letters become illegible—faults arising from some weakness in the currents), the coils should be raised until the signals become distinct.

If this should fail to obtain the desired result, then the antagonistic spring must be weakened; as a rule, the screw D is found sufficient to meet all the requirements of adjustment; and when once A, B, and C have been fixed they rarely require alteration.

D, however, requires to be altered very frequently, and where several stations exist on the same circuit a different adjustment is often required for each.

V. (*a*) The ink-reservoir should never be too full, otherwise the apparatus is apt to become clogged with ink—a result that indicates great carelessness.

(*b*) The communication between the ink-reservoir and well becomes frequently choked with coagulated ink after disuse. This should be cleared with a piece of wire.

(*c*) The ink-reservoir must be frequently cleaned out, and the ink never left in for any length of time. When the day's work is over the paper should be taken out and the instrument should be allowed to run down, to prevent the weakening of the main-spring.

D. Bain's Chemical Marking System.

72. In the previous instruments employed in recording the dot and dash signals, electromagnetism has been used; but Bain, in 1846, devised an instrument by which the same thing was done through the electrolytic effects of the current. Whenever a current passes through an electrolyte, that is, a liquid capable of being decomposed into its constituent parts, the acid element appears at the one pole, and the alkali element at the other. If the liquid be coloured with any

vegetable product, such as red cabbage, its colour will be changed at the two poles. If a piece of paper be soaked in a solution of potassic iodide, iodine will appear at the positive pole, or that in connection with the copper, and potassium at the negative pole, or that in connection with the zinc. The former produces a brown stain upon the paper. If this paper so soaked be drawn beneath a platinum point, a brown line will be drawn upon it as long as the current flows through it in the proper direction. Thus we can form marks by the current and spaces by the absence of the current. Bain did this in the following way (fig. 48). A is a brass drum, whose cir-

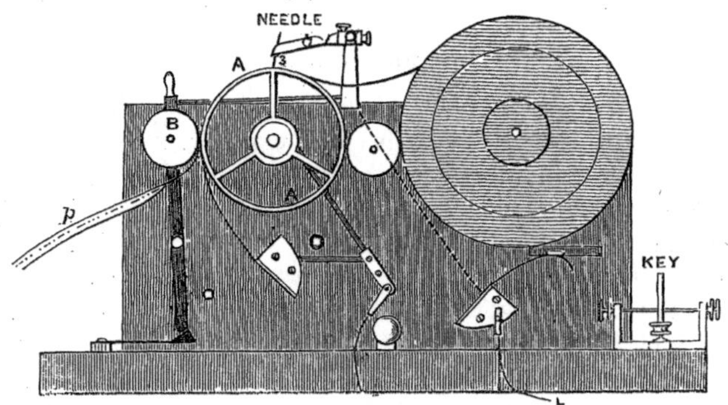

FIG. 48. ⅛th real size.

cumference is tinned, and B is a smaller wooden roller, pressing the paper *p* against it; motion is imparted to this drum by clockwork, so that the paper adhering to it is drawn beneath the piece of wire or style 3, which is maintained in its position by the clip O. The metal point is in connection with the line wire, and the brass drum A is in connection with the earth, so that when a current flows from the line to the earth, iodine appears at the point and leaves a brown line upon the paper. Dots and dashes can thus be made, and we possess all the elements of a Morse telegraph.

73. The solution usually employed in practice is composed of one volume of the saturated solution of potassic ferrocyanide (prussiate of potash), one volume of the saturated solution of ammonic nitrate, and two volumes of water. The ammonic nitrate is a deliquescent salt,[1] and is used to keep the paper damp. The style is of iron or steel wire. When a current flows from the line through paper soaked in this solution it decomposes the electrolyte, the acid radical unites with the iron, and forms *Prussian blue.* Thus dots and dashes can be formed in bright clear blue. The following is a sensitive and useful solution :—

1 part potassic iodide,
20 parts starch paste,
40 „ water.

74. The instrument is worked by a key in precisely the same way as that described for the sounder (§ 59), but it is essential that the direction of the current be attended to, for if it flows in the opposite direction no mark is made upon the paper. Thus a Bain's instrument must always be worked with the copper pole of the battery attached to the front contact of the key.

75. Although relays have been used in connection with this apparatus, the solution can be made sufficiently delicate to be decomposed by the weakest currents. It is not in practical use at present in England, excepting for experimental purposes, for which it is invaluable; but it is so sensitive, and it can register its signals with such marvellous rapidity, that as rapid telegraphy becomes more essential it may probably come again into favour. It was at one time the only form of recording instrument in use in England. It was supplanted by the Morse recorder, which was less liable to get out of order, and which avoided the troublesome operation of preparing paper chemically. In Bain's original instrument a sheet

[1] A deliquescent salt is one which is capable of attracting moisture from the atmosphere and becoming liquid.

of the paper was fixed on a flat horizontal rotating disc of metal, and the metal point moved from the centre to the circumference, so that the dots and dashes were made in a spiral curve. Many ingenious applications of this principle have been attempted by Bakewell, Bonelli, Caselli, and others, but descriptions of such apparatus do not come within the scope of this book.

E. The A B C System.

76. This system, like the needle, is transient or non-recording, but it conveys its signals directly to the receiver by indicating with a pointer the letters of the alphabet arranged consecutively upon a dial. It is the simplest of all forms of telegraphic apparatus for reading messages, but the construction is complicated. The apparatus of this kind in general use in England is Sir Charles Wheatstone's, but there are many other dial forms in use in other countries, viz., Siemens', Breguet's, &c.

77. Wheatstone's ABC dispenses with the use of the battery, and the currents which are employed to move the indicator are produced by the principle of magneto-electricity—one of Faraday's most brilliant discoveries—by which currents are produced by the relative movements of magnets and wires.

We have stated (§ 47) that when a current is flowing through a conductor, the neighbourhood of that conductor is converted into a magnetic field. The converse of this is also true, viz. when a magnetic field is projected through, or traverses a conductor, a current is similarly produced. Moreover, when a conductor traverses a magnetic field a similar effect occurs. Thus to produce these effects motion is necessary, and their magnitude is dependent on the strength of the magnetic field and upon the velocity of the conductor across it.

Let NS (fig. 49) be a powerful fixed bar magnet and C a movable hollow coil wound with a quantity of fine silk-covered wire, whose ends are attached to a galvanometer G.

Let the coil C be rapidly moved over the pole N into the dotted position C′—a powerful momentary current will

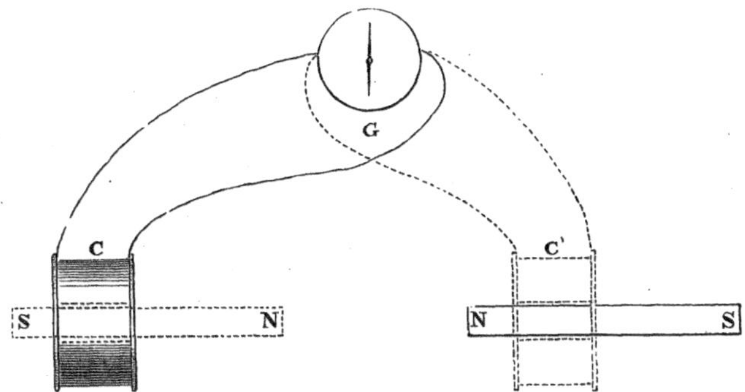

FIG. 49.

traverse the galvanometer. Let the coil be restored rapidly to its original position, C—a current of equal strength, but in the reverse direction, will traverse the galvanometer. Let now the magnet be reversed, and the same movements be repeated, the same effects will be produced, but in the opposite direction. Again, let the coil be fixed and the magnet be movable. If the N pole of the magnet be inserted inside the coil, a powerful current will traverse the galvanometer; and the same will occur, but in the reverse direction, when the magnet is removed. Reverse currents are generated when the poles are reversed. The currents produced by the motion of the coil over the N pole, or by the insertion of the N pole into the coil, are in the same direction as are those produced by the reverse action between the S pole and coil.

78. If the magnet, instead of being a bar magnet, be of the ordinary horseshoe form, and if the coil instead of passing over the end of the magnet simply passes in front of its poles, the same effects occur, though in a somewhat diminished degree; but if the inside of the coil be filled with an iron core this loss is greatly compensated for, because the field is thereby strengthened. Let the coil be moved from C (fig. 50) to C′;

as it approaches N a current is induced in one direction, as it leaves N a current is induced in the reverse direction; as

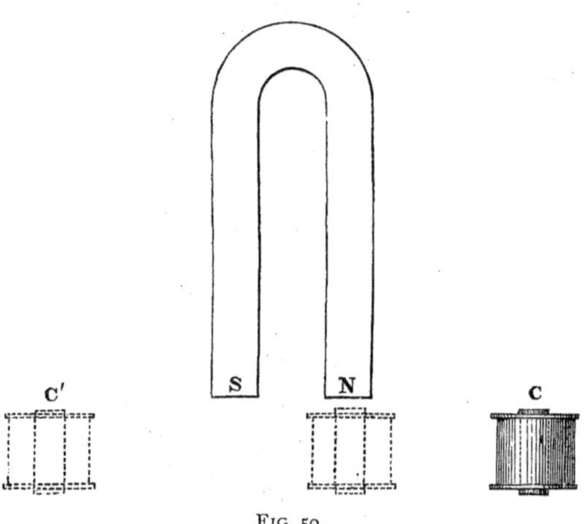

FIG. 50.

it approaches S a current is induced in the same direction as the last, and as it leaves S a current is induced in the same direction as the first.

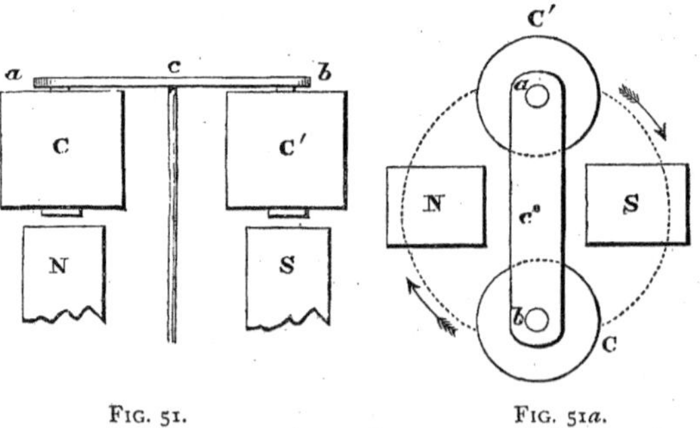

FIG. 51. FIG. 51*a*.

79. Let us take two coils wound like an electromagnet, the two cores being connected by a piece of soft iron,

a b, and rotating about their centre *c*, as shown in section by fig. 51 and in plan by fig. 51*a*; then, as the coil C is approaching N, the coil C′ is approaching S, and currents will be induced in each coil, but in opposite directions. Similarly, when C is leaving N and C′ is leaving S, currents are also induced, but in opposite directions. It is, however, only necessary to connect the wires of each coil together and to the line wires in such a way as to enable the currents induced in the one coil to flow in the same direction and at the same time as the currents in the other, to obtain currents of double strength. Thus by every complete revolution of the coils C C′ four currents are induced.

80. Now instead of making the coils of wire and their iron cores (which are heavy) movable, let us fix the cores and coils to the poles of the permanent magnet, and simply cause the light piece of soft iron, *a b*, to revolve (fig. 52). It is evident that while the cross piece *a b* is in the position shown, the coils being stationary, the magnetic field in which they are fixed will be stationary also; but when the cross piece moves into the position *a′ b′*, then the field is disturbed, its distribution is varied, its strength is concentrated upon the armature, and the coils may be said to have a magnetic field projected through them, which, as has been already remarked, § 77, has precisely the same effect as though the coils themselves were moved across the field. The currents induced by the approach of the cross piece to the soft iron cores are found not to be of the same strength as those induced by its recession, nor are they produced at equal intervals; but by attaching to each pole of the permanent magnet two soft iron cores, as shown in plan by fig. 53 and in elevation by fig. 53*a*, the cross piece is approaching one core while it is leaving the other, and thus the cur-

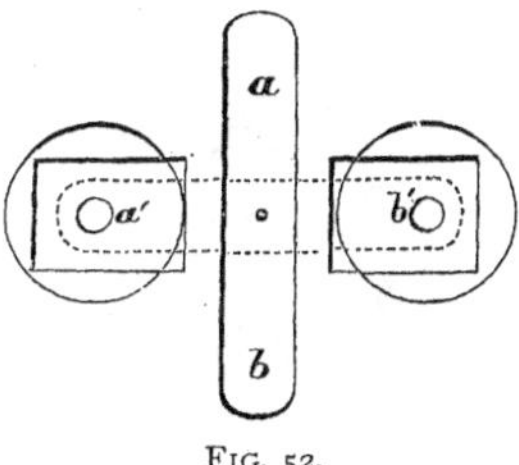

FIG. 52.

rents are equalized in strength and produced at equal intervals. It will be noticed that currents are produced by this method

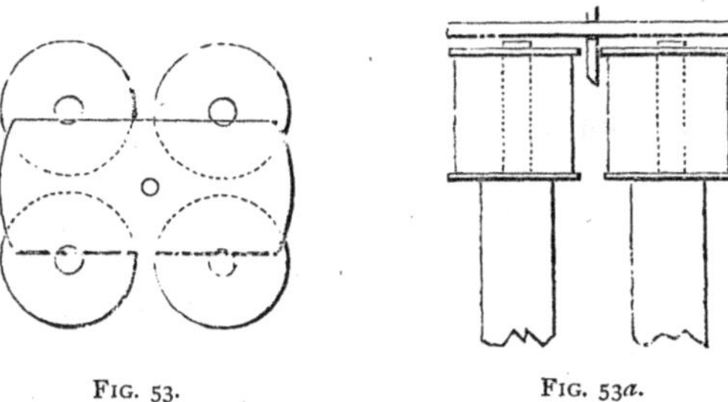

FIG. 53. FIG. 53*a*.

without in any way disturbing the continuity of the circuit, which in fact is never broken.

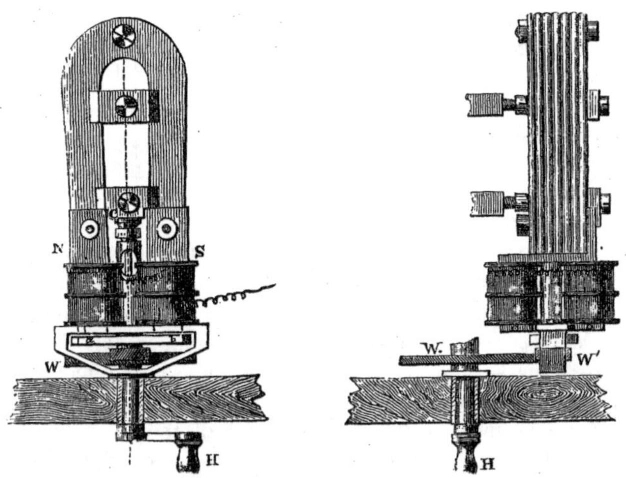

FIG. 54. Bottom view—¼th real size. FIG. 54*a*. Side view—¼th real size.

81. We are now able to comprehend Wheatstone's magneto-electric A B C apparatus: a general view of the portion called the transmitter is shown by figs. 54 and 54*a*.

It is mainly encased in a wooden box, which is not shown in either of these figures.

N S is a compound permanent horseshoe magnet, usually formed of seven simple magnets placed with their like poles together. By means of this arrangement not only is greater magnetism obtained from the same mass of metal, but it is moreover longer retained. To each pole of the magnet two small iron cores wound with insulated wire are fixed, as explained in § 80. These are placed symmetrically with respect to an axis which carries a soft iron armature A B, whose breadth is rather more than the distance between two adjacent cores, as shown above in fig. 53, and which is made to revolve by means of the 'gearing' or driving wheels W W′: these are turned by the handle H. Rotating in connection with the handle is a pointer *p* (fig. 55), which traverses the circumference of the dial: this dial is divided into thirty equal spaces, upon which are marked the twenty-six letters of the alphabet, the three points of punctuation , ; . and a + known

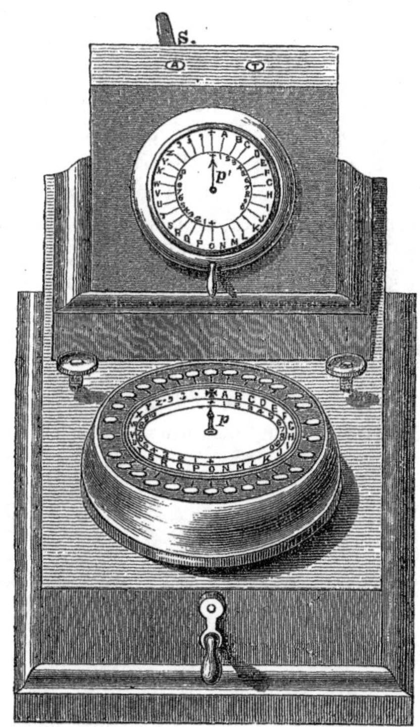

FIG. 55. ¼ real size.

as the *zero* stop: inside these are placed, on each side, the numerals, with the cipher and a +. Opposite to each of the spaces is fixed a key similar to that shown by fig. 56, which can be depressed at will. These keys are placed outside an endless chain, held in position by being passed round a series of small pulleys (fig. 60), and so arranged that only one key can be depressed at a time: the effect of depressing another is to straighten the chain where the first key was depressed, and thereby to throw it up into its normal position.

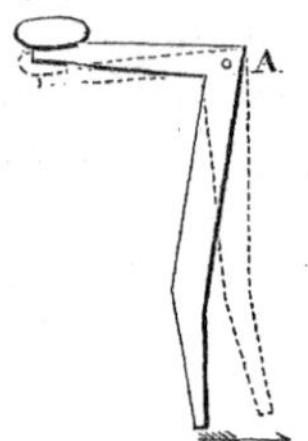

Fig. 56. ½ real size.

If now the handle is turned and the armature sent through one complete revolution in front of the four cores, four separate currents differing alternately in direction are generated. The motion is so adjusted that for each of these currents the pointer moves through one space, and thus for an entire revolution of the armature the pointer goes through four spaces, and four distinct currents are sent in succession along the line to the distant station. When a key is depressed, the motion of the pointer on coming opposite to it is arrested; and the currents, instead of going to line, are cut off. This is effected by means of a carrier arm fixed 'spring-tight' on an axle, and which revolves conjointly with the pointer, but which is thrown out of gear immediately the pointer is arrested by the depressed key: it remains so until this key is raised by the depression of another; and, supposing the handle to be continuously turned, the pointer and carrier arm then resume their march, until again stopped when brought into contact with this latter key. The contact maker K is shown in fig. 57:

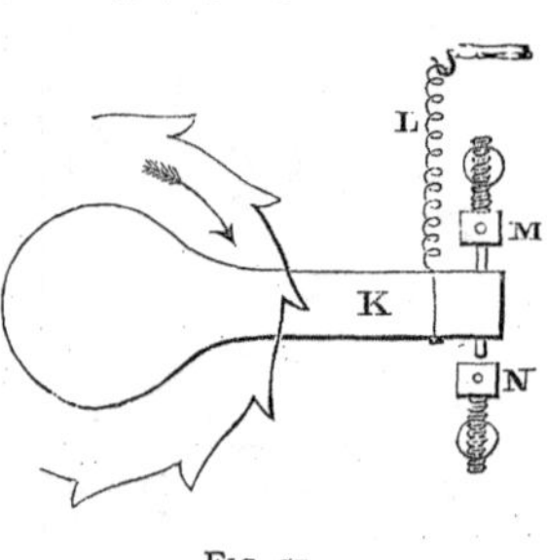

Fig. 57.

L is a spiral spring holding it against the stop M in its normal position of rest. As soon as the handle is turned and the key raised to admit of the carrier arm revolving, K is drawn against N, which is in connection with the line.

82. *The Indicator.*—The dial of the indicator is divided and marked in exactly the same manner as that of the sender. Movable round an axis in its centre is a small pointer *p′* (fig. 55), which indicates whatever letters are sent. The

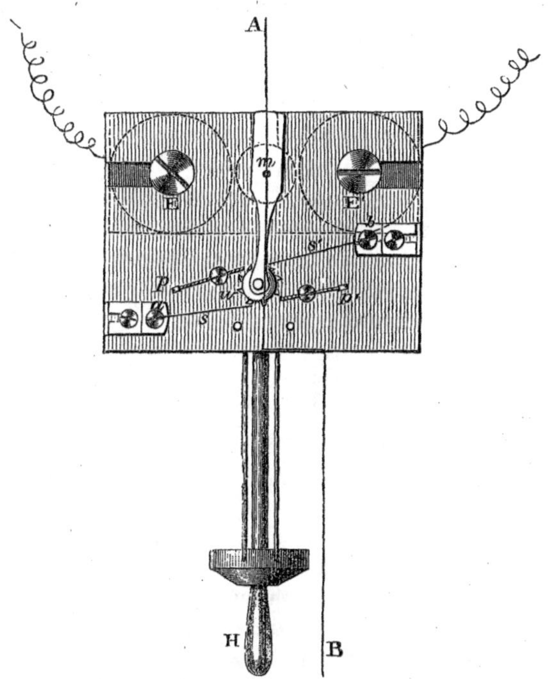

FIG. 58. Full size.

motion of this pointer is regulated by a small escape-wheel *w* (fig. 58), which is propelled by the electromagnetic arrangement shown in plan in fig. 58 and in section in fig. 59. The electromagnet E E′ is wound with fine wire, and is of the ordinary description: between the coils, and lying parallel to the cores, are two small magnets *n s* and *n′ s′* (fig. 59) fixed to an axis, *m m′*, exactly similar in fact to that in Stroh's relay, which has been already described (§ 65). This axis carries the ratchet or escape-wheel *w*, which thereby moves along with it. The mutual attraction and repulsion between the cores, when magnetised by the currents that are sent, and these magnets gives an oscillatory motion to their axis, which causes the escape-wheel to rotate in the following manner. The wheel has fifteen teeth cut on its circumference; its play is regulated by two small pallets *p p′* (fig. 58), and two small steel pallet-springs *s s′*. Each motion to or fro of the magnets forces a tooth against one of the pallet springs which propels the escape-wheel forward through half a tooth, and causes the pointer on the dial to move through one space. A complete revolution, therefore, of the armature in the sender, which, as already remarked, generates four currents, would carry the escape-wheel two teeth forward, and move the pointer through four letters. H (figs. 58 and 59) is an adjusting handle which works the pointer on the dial in the same way as is done by the currents passing through the electromagnet.

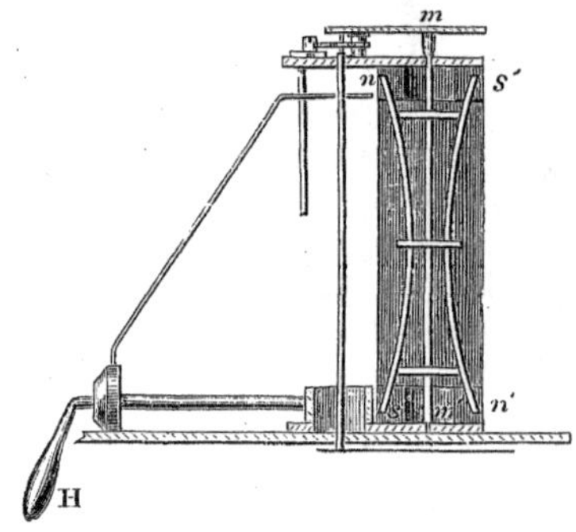

FIG. 59. ½ size.

When two stations are placed in communication with each other, and the apparatus at each is perfectly adjusted, the pointer on the communicator at the sending station moves synchronously with that on the indicator at the receiving station. When at rest both should point to the zero stop, the key of which under these circumstances must be always kept depressed. The effect of leaving all the keys up is frequently to disconnect the circuit.

83. When there are only two stations in circuit Wheatstone's A B C is found to work very satisfactorily indeed. The addition of every intermediate station introduces an element of danger, and complicates to a great extent the adjustment of the apparatus. Four stations fitted with these instruments upon the same wire may be accepted as the limit of safety: only under quite exceptional circumstances should five be tried. As these instruments are invariably employed either upon circuits over which comparatively little work passes or for private wires, an alarum bell is used in connection with them for the purpose of drawing attention when any communication is to be sent. This bell can be cut out of circuit by the movement of the switch S, the top of which is shown in fig. 55. When this switch is at A the alarum and indicator are both in circuit, when turned to T the alarum is cut out, and the indicator only is in circuit.

84. The adjustment, more especially of the indicator, is a delicate matter, and requires a considerable amount of skill and training before it can be undertaken with safety. If the pointer in the indicator jumps, or moves on in advance of the letters sent, the currents are either too strong or the pointer is too lighrly adjusted. Either the armature in the sender should then be moved farther back from the cores, or the play of the escape-wheel in the indicator should be lessened by tightening the small screws and springs. The former are provided with split heads in the usual way: the screws *a* and *b* (fig. 58) regulate the tension of the latter.

If on the other hand the indicator pointer lags behind and drops letters, the currents sent are too weak, or the springs are too stiffly adjusted. Either the armature should then be approached to the cores in the sender, or the play of the ratchet-wheel in the receiver should be assisted by easing the studs and springs.

It occasionally happens that the endless chain in the communicator, by means of which the motion of the keys is regulated, is either stretched to such an extent that more than one key can be depressed at the same time, or it becomes contracted so as to prevent even one key from being depressed. In the first case the chain requires to be tightened, and in the second to be slackened. Provision is made for effecting this by means of the arrangement shown in fig. 60.

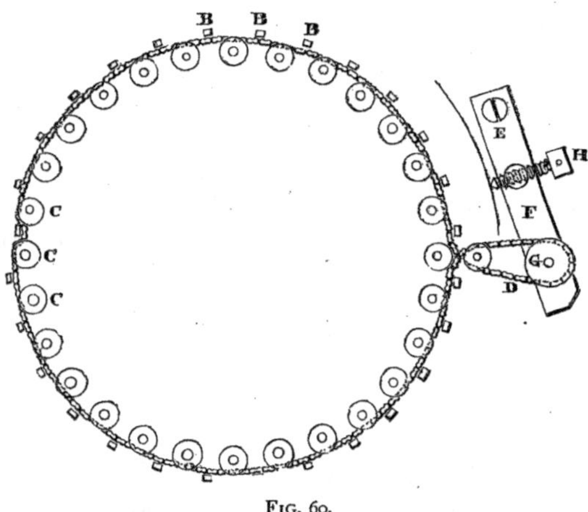

FIG. 60.

The endless chain is passed around an additional pulley G, fixed upon a lever F, pivoted at E. In connection with this lever is an adjusting screw H. By screwing H in, the lever is drawn outwards, and a greater portion of the chain being thus taken up in connection with it, there is less slack left. By unscrewing H, on the other hand, a portion of the chain is released and the length available for the action of the keys may be thereby increased to whatever extent is desired.

85. The main difficulties to be encountered with these instruments are due to atmospheric electricity. Lightning frequently deranges them to a great extent. Not only does it readily fuse the coils, on account of the wire with which they are wound being necessarily so fine, but by demagnetising or even reversing the polarity of the small magnets in the indicator, it interferes with their play and renders a fresh

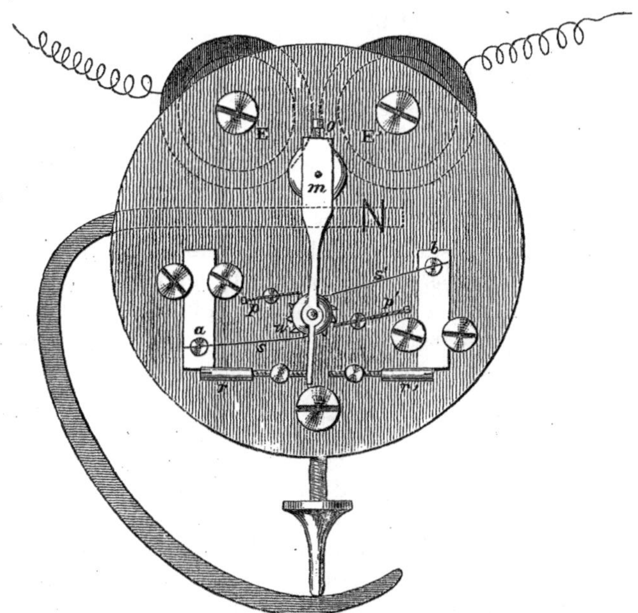

FIG. 61. Full size.

adjustment or remagnetisation necessary. This latter danger has been overcome to a great extent in the form of indicator which is now issued by adopting the principle which was introduced into the coils for needle instruments (§ 50), that is to say by employing induced instead of permanent magnets. Fig. 61 shows in plan the latest form of indicator. The escape-wheel and its adjustment are almost exactly the same as in the earlier issue: two additional screws *r* and *r'* are added by means of which the play of *m* can be better regulated. The compound magnet shown in 59 is now dispensed with, and in its place two soft iron armatures connected by a small rod pivoted into *m* are employed. The upper of these is shown at O. These soft iron armatures are kept in a magnetised condition by means of the large horse-shoe magnet N S partly shown in the figure: the S pole is at the lower end of the coils. The same beneficial results attend this arrangement as have been already referred to in the single needle instrument.

Occasionally, too, in the case of a heavy thunderstorm, the large permanent magnets in the sender have their magnetism reduced: the currents generated by them become weak for the adjustment to which the apparatus has been set, and another has to be substituted: instances, too, have occurred, in England even, where the lightning has removed every trace of magnetism from these large magnets; they are very rare, however, and quite exceptional.

F. Type-printing Instruments.

86. These are instruments which record the messages sent in bold clear Roman type. Many ingenious forms of apparatus have been devised and practically used for this purpose, but only one has attained any considerable employment in Europe. This is Hughes'. It is shown by fig. 62. It differs from all others in being principally mechanical. Only one current of short duration is employed to register each letter. The instruments at the sending and the forwarding stations are identical in construction and movement. Their type wheels a (fig. 62) having the letters of the alphabet raised on their peripheries, and attendant apparatus are kept rotating synchronously and simultaneously. The sending apparatus is like the key-board of a piano, with the letters of the alphabet and

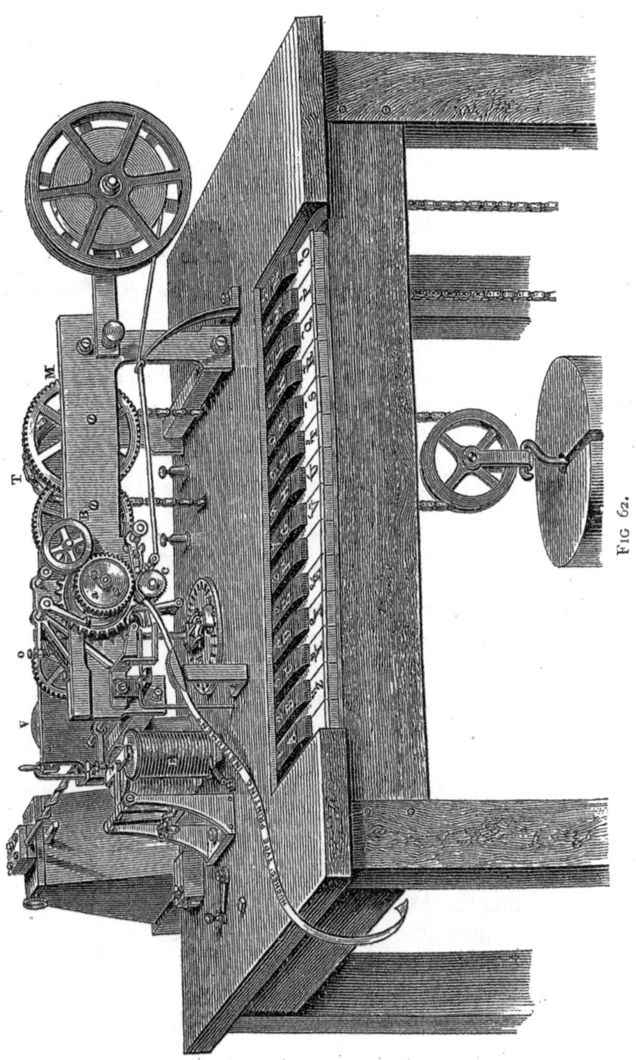

Fig 62.

any other signals needed engraved on the keys; when one of these keys is depressed a pin is raised, which just catches a '*chariot*' (shown at A in the figure) rotating with the type wheel, and thus sends a current to the distant station. This current causes the paper at both stations to be lifted at the same time into contact with the type wheels. Both wheels having their circumferences coated with printers' ink and rotating in unison, print the letter corresponding to the pin raised at the sending station. The same movement causes the paper to be moved forward a space ready for the next signal. In this way, by touching each key required successively, words and sentences are spelt out, spaced, sent and recorded at both stations simultaneously. Fig. 63 gives a sample of a short sentence so printed.

ALLS WELL THAT ENDS WELL .

FIG. 63.

87. The mechanical construction of the apparatus is exceedingly ingenious and perfect; but as it is in use only to a limited extent in England, a full description of it does not fall within the scope of this work.

88. The electrical arrangement is exceedingly simple, sensitive, and novel. The current which is sent does not *attract* an armature, but it so weakens the polarity of an induced magnet as to cause it to *release* an armature which is thereby pulled away by the tension of a powerful antagonistic spring. The armature is restored to its normal position by the mechanical action of the instrument. This electrical arrangement is indicated by the following figure (fig. 64). N S is a powerful permanent steel magnet, having two soft iron pole pieces, to which two soft iron cores are permanently attached, surrounded with coils of wire which form part of the line wire: *a* is a movable soft iron armature and *s*

an antagonistic spring. When this armature is placed upon the pole pieces, it is held there by the magnetism induced in the pole pieces by the permanent magnet N S, and it will bear a considerable tension of the spring *s* before it will be torn off; but if a current passes through the coils in such a direction as to induce in the cores a polarity the reverse of that induced by the magnet, the armature will be released and it will fly back with the full force of the tension of the spring. The instrument thus becomes sensible to exceedingly delicate currents, and it makes its signals with the full mechanical force of a strong spring.

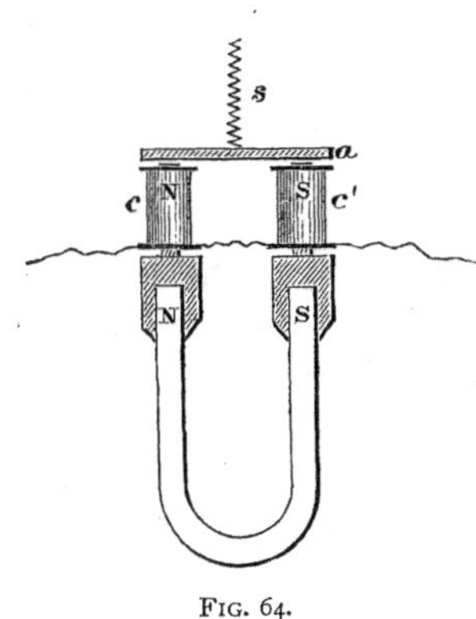

FIG. 64.

89. It is to be regretted that this beautiful instrument has not met with greater favour in England. It is very much used on the Continent, especially in France, where it is a great favourite. It is, however, expensive both in its prime cost and in the character of the labour required to work it; for though, as a rule, it only requires one clerk at each end, and dispenses with writing altogether, this clerk must be of the most intelligent and experienced class, and therefore must be comparatively highly paid.

G. A COMPARISON.

90. The instruments which have been described in the previous pages have shaken themselves out as it were from a mass of beautiful apparatus of the same species, which have been practically tried in England, and each has in its own sphere proved itself to be the best adapted to the purposes which it was intended to serve. This struggle for existence is, however, still progressing, and it is quite im-

possible to say that each is an example of the survival of the fittest, for the probability is that one or other of the forms now in use will eventually be jostled out of employment by its more perfect competitor.

91. The Hughes is but very little used. The Bain is merely a modification of the Morse, and the Bell is gradually being replaced by the Sounder or the Needle. In drawing a comparison of the relative advantages of the different instruments used, these three instruments, the Hughes, the Bain, and the Bell, will not be considered, though each merits better treatment and deserves more consideration than it has received.

In making this comparison we will take into consideration their simplicity in construction and working, their economy, their rate of working, and their special adaptation for the purposes they are each peculiarly qualified to fulfil.

1. *Simplicity in Construction.*

92. Simplicity in construction is a great desideratum, especially in all uncivilized countries; and even in civilized countries, when the operators are unskilled or ill-educated. Of all the different instruments described the direct Sounder is unquestionably the simplest in construction; but when it has the relay added to it, it is difficult to say whether it is simpler than the Needle. The Needle has this advantage over the Sounder or any other instrument: it has no points requiring adjustment, and though the arrangement of the sending portion of the apparatus is more complicated than the simple key of the Morse and Sounder, the receiving portion is much more simple than the relay and Sounder. The Morse involves a complicated and expensive trainwork of mechanism and a governor to maintain the paper in uniform motion; and the inkwriter demands a special contrivance to keep up the flow of ink; the inking disc is independently rotated in a direction opposite to that of the paper. Both Sounder and Morse have galvanometers at-

tached to them, and though these are scarcely needed with the former instrument, they are of great convenience in noting the sending currents.

The A B C is both complicated and costly in its construction, and in the simplicity of its parts cannot compare with its competitors.

Hence, as generally employed, the Sounder is unquestionably the simplest form of apparatus constructed for telegraphic purposes.

2. *Simplicity in Working.*

93. The A B C involves no technical skill in sending and receiving messages. An hour's practice will enable any child or old person to send or receive a message by its means. It is only necessary to watch the movements and pauses of the indicator to read, and to follow the letters of the alphabet to send. The Needle instrument requires no special skill with the hands to send, though rapidity of sending is acquired with practice only; but the Morse recorder and Sounder involve technical skill, long practice, and experience both to read and to send. In the case of the Sounder, however, if once a person learn, or be taught to read by sound, sending becomes not only comparatively easy, but remarkably accurate. The A B C instrument is therefore the simplest in working, and the Morse is the most complicated; the Needle and Sounder may be bracketed together, for the ear is probably as apt a pupil in reading as the eye, and the hands in sending follow whichever organ is employed in receiving. The Sounder has this immense advantage over the Needle, that it allows the receiver to concentrate his eye upon the form on which he writes the message he is receiving, while the needle-clerk has to glance alternately from his needle to his paper, and thus perform two operations to one performed on the Sounder. Thus the Sounder is not only as accurate, but is far more rapid than the Needle. But special contrivances are needed to concen-

trate the sound, and it is liable to be abused by eaves-droppers.

3. *Economy.*

94. The prices of the different instruments, and the cost of their annual maintenance, are shown in the following table :

	Prime Cost.			Annual Maintenance.		
	£	s.	d.	£	s.	d.
A B C	25	0	0	3	10	0
Needle	7	10	0	2	10	0
Morse	25	0	0	7	10	0
Sounder	12	0	0	4	10	0

4. *Rate of Working.*

95. The useful speed of a non-recording telegraph instrument is in reality limited by the rate at which a clerk can write; but in recording instruments this is not so, because if one clerk cannot write as fast as the instrument records, a second clerk can be appointed to follow him. All these instruments which have been described are, however, limited in speed by the rate at which a clerk can send or manipulate his key. The Single Needle in expert hands frequently attains 35 words per minute; and the average rate at which an ordinary needle circuit works is 25 words per minute. The Morse inkwriter, under the same circumstances, attains 40 words per minute, and the average rate at which an ordinary Morse circuit works is 35 words per minute. Very expert manipulators sometimes attain as many as 20 words per minute on the A B C, but the ordinary rate of working with this form of instrument rarely reaches 10 words per minute, and the average does not exceed 5 words per minute.

The Sounder attains the same speed as the Morse, and practically can be read faster than any clerk can write; but there is no advantage in exceeding this speed except in conversation, and therefore for all ordinary purposes the Sounder attains a rate of working in experienced hands of from 35 to 40 words per minute. But the number of words per minute which an instrument can transmit is no criterion of its value as a fast-working apparatus. We must regard the total work it can do in a day—the number of messages it will transmit without unnecessary delay. Now the nature of business is such in England, that the chief bulk of the messages are sent between the hours of ten A.M. and one P.M., and it is essential that between these hours the wires shall not be overcrowded with messages. The following table may be taken as giving a fair average of the number of messages which each instrument transmits in an hour and in a day :—

	Hour	Day[1]
Sounder	60	250
Morse	45	175
Needle	30	125
A B C	15	60

Thus the Sounder is by far the fastest instrument, and a day's work on a Sounder will exceed that on any other instrument similarly worked. Of course this rate of working depends upon the number of words which each message contains. In England the average is 34 words, including the ordinary service signals, and each word averages 4·5 letters. The reason why the Sounder is so much quicker than any other instrument is, that as both stations are simultaneously sending and receiving, corrections are made at once, the receiving clerk keeps up with the sender, and there is never any waiting for repetitions or acknowledgements. A clerk receiving a message by means of the Sounder confines his eye to what he commits to paper—his mind is free to follow the sense of the message. He is simply in the position of being addressed by a clerk, perhaps hundreds

[1] This does not indicate what the instruments can do, but what they *do* do in a day with English telegrams.

of miles away, who dictates each word not as it is spoken but as it is spelt. The ear does not tire like the eye, hence the clerk can maintain the rate of working for longer periods than with the recorder or needle.

5. *Special Adaptability.*

96. Telegraph instruments are required for many different purposes, and are placed in many different situations. They are required for the transactions of the ordinary business wants of the country and of the domestic relations of the community. The transmission of that enormous mass of news that now forms such a large portion of many newspapers has to be performed by them. They are necessary for the various purposes of the railway companies in regulating the traffic and moving the trains upon their railways. They are employed between the mansion and the stables, between the merchant and his counting-office, between the shop and the parlour. They are worked by highly trained and well paid manipulators, by inexperienced and insufficiently paid boys and girls, by the assistants of the flourishing tradesman, and by the superannuated village grocer or his wife. Thus each instrument has its special sphere.

The A B C is specially adapted for private wires and for small village post-offices, where messages are few and far between, and where skilled and trained labour can neither be found nor paid for.

The Needle is specially adapted for railway purposes and for linking together several towns on one wire, neither of which singly does much work, but where all together occupy a wire. No instrument that has ever been devised so fully meets the requirements of a railway. Its manipulation is easily learnt, and not easily forgotten. The apparatus never wants attention, and is always in order. Many more stations can be grouped together on one wire by means of it than by any other instrument.

The Morse and Sounder are specially adapted for news

and for commercial purposes where the amount of business is sufficient to justify the employment of skilled labour; and of these two the Sounder is unquestionably the superior, and will ultimately supersede the other.

CHAPTER IV.

CIRCUITS.

97. We have defined the circuit (§ 10) to be the whole path along which the electricity is supposed to flow; and we may consider the two cases in which the currents are flowing and in which they are not flowing. If we construct a wire between A and B (fig. 65), the end at each place

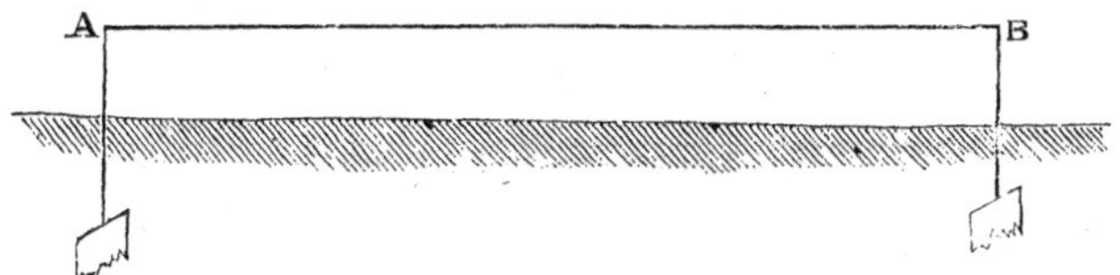

Fig. 65.

being connected with the earth, then the whole path from A to B along the wire and back again from B to A through the earth is the circuit; and this circuit may either have a current flowing through it, or it may be free from all current. In the latter case the circuit is said to be *open*, in the former case to be *closed*. There is a second condition of a circuit in which no current is flowing, called the *equilibrium* circuit, because every point of it is kept at the same potential by having similar poles of different batteries of equal electromotive force attached to it; but this is a form of circuit not used practically for telegraphic purposes.

These three conditions of the circuit are shown in figs. 66, 67, and 68 respectively.

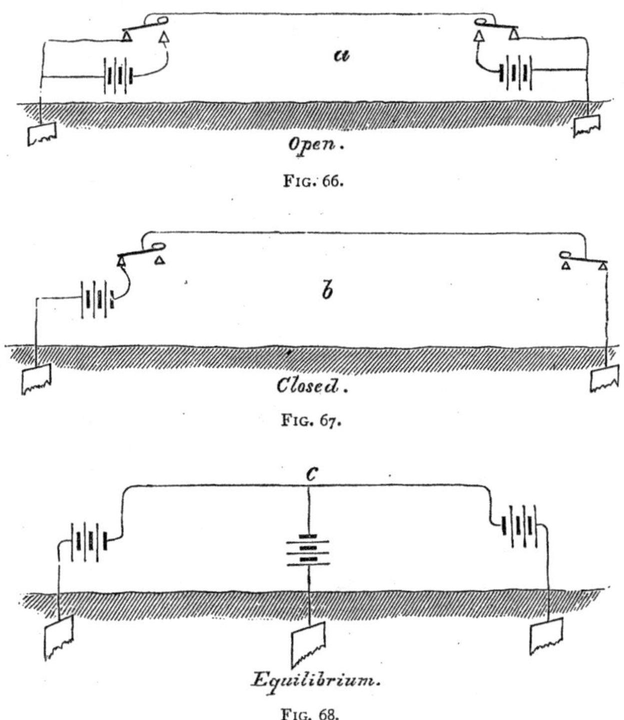

FIG. 66.

FIG. 67.

FIG. 68.

98. The earth is considered simply as a part of the circuit, offering a certain resistance to the flow of electricity through it, but this resistance owing to its infinite dimensions is practically nothing. Hence the earth may be said to allow currents to flow through it in any direction, and without any obstruction or interference when considered as a whole; but in dry and rocky formations, and for limited distances, it does offer resistance, which can be measured, and which

introduces disturbances that have to be eliminated or allowed for.

The battery which generates the current, and the apparatus which renders it evident to the senses, are essential and important parts of the circuit, and their resistance is material in determining its good working.

99. Thus we see that whatever is in the path of the current—whether it be in the battery itself, in the apparatus, in the line wire, or in the earth—whatever, in fact, offers any resistance to the passage of the electricity, is *the circuit*, and this circuit, for telegraphic purposes, may be either open or closed.

100. The needle instrument is invariably worked on the open circuit system. The normal position of the needle when at rest being in the vertical implies the absence of current, and the motions to the right and left, due to the reversal of currents, imply some re-arrangement of the circuit resulting in this reversal. K and K′ (fig. 69) are the

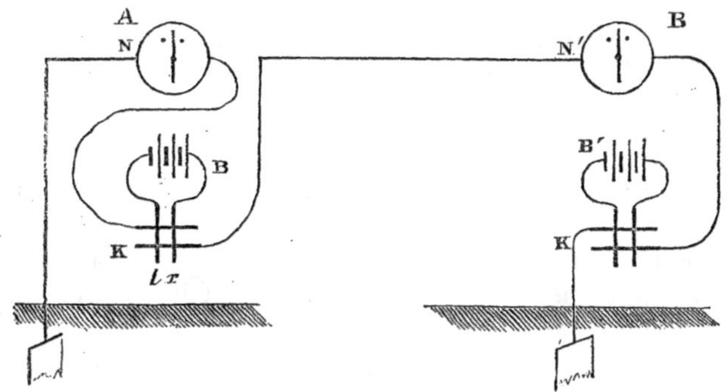

Fig. 69.

commutators, B and B′ are the batteries, N and N′ are the coils and needles. The action of the commutator is described in § 48. When *r* of A is depressed the current flows from station A to station B, when *l* is depressed the current flows in the reverse direction. The needle deflects in the direction of the current. Thus at A we can make the needle at B deflect at will in either direction by depressing either *l* or *r*; and similarly at B we can make the needle at A deflect in either direction at will.

101. The Morse or Sounder system is also in England invariably worked on the open circuit system, but it is worked with or without a relay. Open circuit working without a relay is called *line current working.* It is shown in fig. 70. B is the battery, which is usually connected up

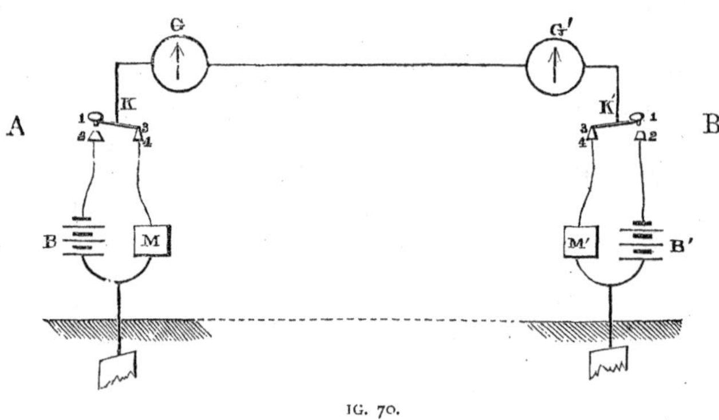

IG. 70.

with its zinc pole directed to line; this arrangement is found to cause less disturbance when faults exist, especially in underground and cable portions, than when the copper pole is directed to line. K is the key described in § 59, which on depression at A causes the currents to flow. G is a galvanometer which allows the existence of these currents to be seen, and M is the recorder or sounder worked by the currents received from the distant station. Now when station A wishes to communicate with station B, he depresses his key K, 1 is brought into contact with 2, which thus brings the battery B into action, sets a current flowing, causes the galvanometer G, as well as the galvanometer G' at B, to move, and works the recorder M′ at B, his own recorder not being affected. Thus if the attention of B is not attracted by the sound or motion of the recorder, it is by the galvanometer, and A knows by his own galvanometer whether his currents are flowing properly or not.

102. Line current working is only used for comparatively short lines, except in the case where direct ink-writers are used (§ 69), which, dispensing with the heavy work of embossers, need but weak currents, and are therefore used upon lines of about 100 miles in length or under. Where, however, the distance exceeds this length, where the insulation of the wires is indifferent, and where abnormal resistance is introduced through the insertion into the circuit of intermediate or wayside stations, relays become necessary. We then work by *local currents*. M, fig. 71, is

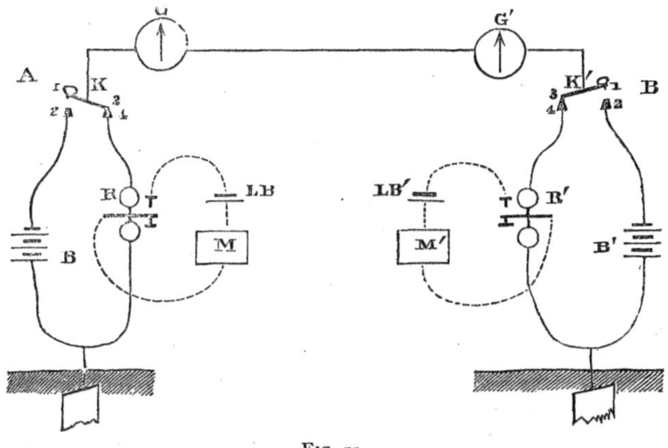

FIG. 71.

the recorder or sounder, L B a local battery, R the relay which takes the place of the recorder in fig. 70; all the other connections are the same. When the key K at station A is depressed, 1 is brought into contact with 2, a current flows through 1, 2, G, G′, K′, 3, 4, R′, to earth and back to B the battery at A. In flowing through R′ it moves the tongue of the relay and completes the local circuit by which the local current flows from L′ B′ through M′, and records its signals.

103. Allusion has been made to the effect of the introduction of intermediate instruments. The introduction of such apparatus in no way affects the theoretical working of the circuit. If in either of the above two cases the earth wire at B instead of being carried direct to earth were attached to another line wire extending beyond it, it would be seen that the circuit would still remain whole and open, and when any one station worked every other station would be affected. The connections at an intermediate station on a circuit working with local currents are shown in

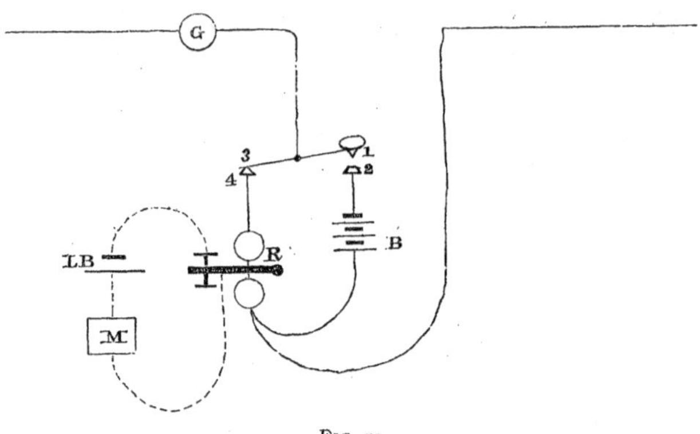

FIG. 72.

fig. 72. It is evident that while the apparatus is idle the continuity of the circuit is maintained through 3 and 4 of the key, and that when the key is depressed the currents flow through both the up and down line without affecting the recorder at the intermediate station itself, but operating those at the other stations. If simple line working be in use it is only necessary to replace R by M alone, and remove L B. The mode of connecting up a needle instrument intermediate is symbolically shown in fig. 73, where for variety drop-handle instruments are shown.

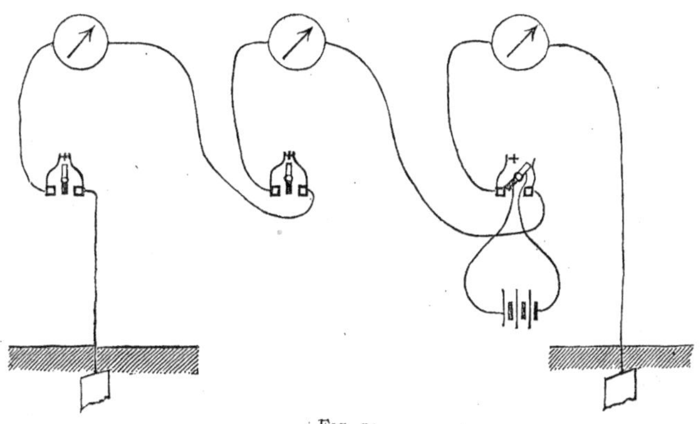

FIG. 73.

104. The closed circuit system has never been a favourite in England. It has been frequently tried, but owing to the greater consumption of materials in the battery compared with the consumption on the open circuit, and to the detrimental action of continuous currents on the gutta-percha covered wire, so much used in England, it has as frequently been abandoned. It is shown in its simplest form in fig. 74. M is the recorder, K the key, B the battery, as before; but there is only one battery at one station, and not a battery at each station, as in open current working. The key K has a movable handle or *switch*, as it is called, which normally is closed, as shown in K′ at station B, and connects the battery to line, so that when the circuit is idle the current is flowing. If A wishes to communicate with B this switch is pushed aside, the current ceases to flow, the circuit is open, and A works, as in the open circuit system, closing the switch when he has done working. If B wishes to communicate with A he also opens the switch; but when he depresses his key he simply does so to complete the circuit for A's battery, and

therefore he works the circuit by means of that battery. A large number of intermediate stations can be inserted on such

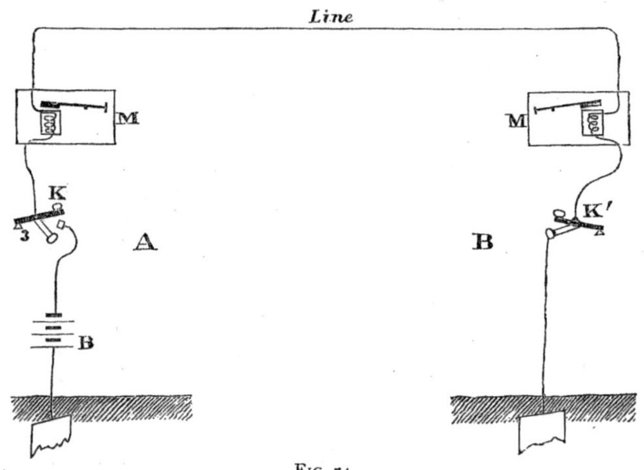

Fig. 74.

a circuit, and it is evident that if they all keep their switches closed the current flows throughout the whole circuit; and any station, by opening his switch, can break in and operate every instrument upon the circuit by opening and closing the circuit of the one battery fixed at one of the terminal stations.

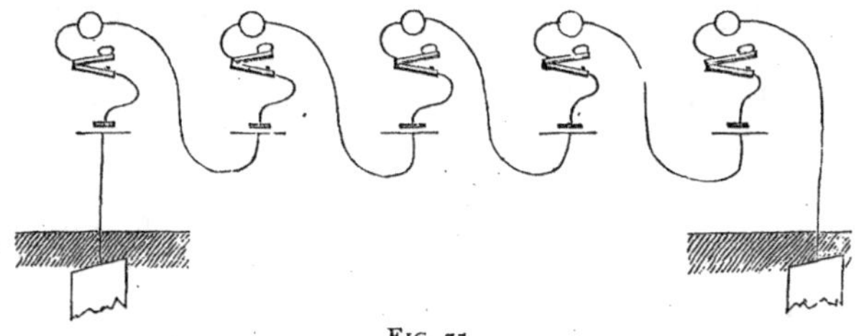

Fig. 75.

105. Closed circuit working is very generally adopted in America, in Australia, and other colonies. It is also much

used in Germany, where some circuits have as many as fourteen stations upon them. As, however, a battery at one end of such a circuit, if it be long, has its currents very unequally distributed amongst the different stations owing to the effects of leakage, each station is sometimes provided with a part of the battery, which forms part of the circuit. This is shown in fig. 75.

106. There can be no doubt that where many intermediate stations are fixed on one wire worked on the Morse principle, the closed circuit system offers considerable advantages over the open circuit system; for the inconveniences arising from the difficulty in maintaining the accurate adjustment of the apparatus, when working to different stations at different distances, owing to the variations in the current, is to a considerable extent avoided. The current at the same station is constant. But in England we never do use the Morse on such circuits. The Needle is far preferable, and in that system no adjustment whatever is needed. It is an exceedingly rare thing to fix more than four stations on one Morse circuit, for the simple reason that it is almost impossible to group four Morse stations together without filling the wire—that is, without producing a sufficient number of messages to occupy the wire during the whole of the day. If there is not enough work to fill the wire, the needle instrument, from its simplicity, economy, and certainty, is used in preference. So that the necessity for employing the closed circuit system in England has not been experienced, while its extra cost, as compared with the open circuit and injurious action on gutta-percha-covered wires, have proved it to be really objectionable. We mention these facts because surprise is often expressed by Colonists and Americans that the closed circuit system is not used in England. Every country has developed its own system, and the conditions which have rendered the closed circuit necessary in America do not exist in England.

107. There is another mode of working a closed circuit,

which was originally introduced in America, has been used on the Hanoverian lines, and is now being applied to the State Railways of India. Instead of breaking the closed circuit by a switch, and converting it really into an open circuit with the key worked at any point, the instruments are caused to work by the interruption of the current. Relays are used which are held in their normal position of rest by the current, and which are caused to complete the local circuit by the interruption or cessation of the current. It is not much used.

108. When any length of gutta-percha wire, either in a submarine cable or in underground pipes, forms part of a circuit, it tends to diminish the speed of working by accumulating upon the surface of the wire a portion of the current which otherwise would proceed to the distant station to record or register its marks. A similar effect, but on a smaller scale, occurs on overground wires. When, however, such lines are very long, this effect of induction, as it is called, becomes evident. To overcome this abstraction of the currents, slower and more deliberate sending is necessary. The key must be held down longer to allow a dot to be made, for the short and smart dots made upon a short aerial line are entirely lost on an underground, submarine, or a long overground circuit. The loss of speed is to a large extent remedied by *double current working*. A second current, reverse in direction to the first current, and sent immediately after it, not only hastens the discharge or clearance of the wire of the charge accumulated upon it, but it enables the relays to be worked in their most sensitive or delicate, and, therefore, most rapid position. This method of working is shown in fig. 76. G is the galvanometer, K the key, R the relay, and M the recorder or sounder; but in addition to the ordinary line battery B there is a second battery B′, whose opposite pole to that of B is connected with earth, while its other pole is attached to the point *a′* of a switch S, whose handle is movable at will from *a* to *a′*.

The switch-handle itself is in connection with the back contact of the key K; and in its normal position it is in contact with *a*, and it thus enables the line currents received from

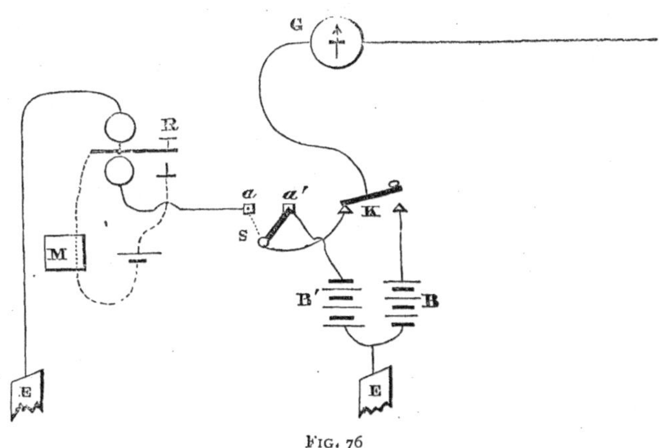

FIG. 76

the distant station to pass through the relay R, which is in connection with *a*, and work the instrument M. When it is desired to send, the switch-handle is moved to *a'*, which allows the current from B' to go to line. This current passes through the relay at the distant station, and holds its tongue away from its contact points; hence all antagonistic springs or other forces are dispensed with, and the relay is in its most sensitive condition to respond to any change in the current. When the key is depressed this current ceases, and a reverse current is sent into the line, which moves the tongue of the relay in the proper direction, and marks are made. Thus there are always currents flowing when the switch is turned for sending, and it is the reversal of these currents which works the apparatus. Polarised relays are necessarily employed with this system of working, for they alone are obedient to the changes in the direction of the current.

109. The double current system of working not only expedites the rate of working on submarine, subterranean, and long overground circuits, but it frequently enables the working of the circuits to be continued in the face of considerable interferences and disturbances inherent to overground wires. It deprives relays of any effects of residual magnetism; it allows circuits to be worked with less powerful currents, and, consequently, it enables circuits to be worked to much greater distances. When polarized relays are used with single currents only, the antagonistic force of magnetic attraction requires that the tongue of the relay be kept further from the core attracting it for working purposes. It is for this reason that under such circumstances non-polarized relays are found to be more sensitive. But when double current working is resorted to the tongue is placed in its most sensitive position, and all antagonistic forces are removed.

110. It is objected to this system of working that messages once commenced cannot be interrupted for corrections or enquiry. The practice is to turn the switch every now and then to see if all is going right. Twenty years' experience of this system have shown that the objection is groundless, and only theoretical.

The difficulties of adjustment inherent to open circuit working with single currents, owing to the variations in the strength of the currents received from different stations, are entirely overcome in double current working, for whatever be the variations in the prime current the reversing current is equally and similarly affected, and thus the moving force and the antagonistic force vary together, and are self-adjusting. Double current working is therefore almost invariably adopted on Morse and Sounder circuits having intermediate stations upon them.

111. The strength of the currents working circuits diminishes with their length, not only in consequence of the additional resistance, but from the injurious effects of climate and weather upon the wire and their supports. There is a

limit to the increase of battery power. It is not safe to use more than 120 cells. It is therefore difficult to maintain uninterrupted communication in England for distances of over 400 miles. In dry climates, and where purely aerial wires are used, much greater distances are possible; but in all countries a distance is at last reached where direct working is impossible, and where it becomes necessary either to repeat the messages themselves by clerks, or to introduce *mechanical repeaters* or *translators* at some intermediate station to bring into play fresh currents, and, therefore, renewed strength. By this means it is possible to work to any distance. Thus the Indo-European line from London to Teheran, a distance of 3,800 miles, is worked directly, without any retransmission by hand, by means of five repeaters.

The connections of a repeater are shown in fig. 77.

112. The principle consists simply in converting the lever of the recorder or sounder M or M′ into a key which is moved by the attraction of its armature between two contact pieces corresponding to the front and back contact pieces of the key. The electromagnet of the recorder thus replaces the hands, and the motions of the key at the distant sending stations are thus repeated at the translating station. This automatic key brings into play a fresh line battery B, which sends on a fresh current to the distant receiving station. Let us fix ourselves at the repeating station, where there are two sets of identically similar apparatus, as shown in the diagram, and imagine that the up station A is sending to the down station B. The currents from the up line enter the lever *t′* of the recorder M′, and pass by 1 to R, the relay of the up set of apparatus, which they work, they then pass to the earth plate E, and return by the earth to A. The tongue of the relay R moves from 3 to 4; it thus completes the local circuit of the local battery L B, the recorder M works, its armature is attracted, the lever rises from 5 to 6, and the battery B sends currents viâ 6 and *t* to the down line. These currents correspond in duration with those received from A.

Next let us assume that the down station B is working to the up station A. The currents from the down line enter

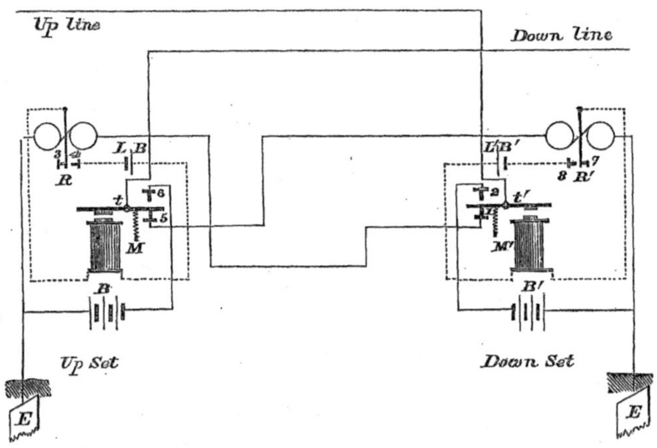

Fig. 77.

the lever *t* of the recorder M, and pass by 5 to R′ the relay of the down set of apparatus, which they work, pass to the earth plate E, and return by the earth to B. The tongue of the relay R′ moves from 7 to 8; it thus completes the local circuit of the local battery L′ B′, the recorder M′ works, its armature is attracted, the lever rises from 1 to 2, and the battery B′ sends currents viâ 2 and *t′* to the up line. These currents correspond in duration with those received from B.

113. In practice the connections are not so simple as those shown in fig. 77. Galvanometers are used on each line wire to show if the currents pass correctly. Hand keys are used, which can be thrown into both up and down circuits by means of switches, so that the circuit can be divided, and the repeating station can work separately, either to A or to B, without translation. Again, when the double

current system, and other fast systems of telegraphy are used, the connections become much more complicated, and special automatic switches are employed to maintain the reverse currents. The recorders or sounders are replaced by relays, and many ingenious contrivances are added to facilitate adjustment and to secure good working. But the principle remains the same in each case, and is that illustrated in fig. 77.

114. There is, however, a limit to the number of repeaters which can be employed on one line. The motion, friction, and inertia, both magnetic and mechanical, of the moving parts and the introduction of disturbing electrical causes, prevent the duration of the contact of the tongue of the relay from being the exact counterpart of that of the sending key. It is of less duration. Retardation therefore takes place, and the rate of working is reduced with each relay added. In no case in England do we introduce more than one repeater; and when that repeater is placed at the end of a submarine cable, like that connecting Ireland and Wales, an actual and decided increase of speed is obtained, due to the fact that the speed of working of the whole circuit is made that of its longest disturbing section alone: if working without the repeater other retarding influences are superposed upon that of the longest section. Thus a repeater at Haverfordwest has increased the rate of working between London and Cork about ten words per minute.

115. We can scarcely conclude this brief description of the mode of joining up instruments in circuit without referring to the circuit arrangements required to serve a portion of the country. We have taken the Isle of Wight. The diagram, fig. 78, shows how all the villages and towns are connected together and with their great centres of communication, Southampton and London. It illustrates also the way in which the different instruments are employed. Thus little places like Niton, Sea View, St. Helens, Brading, which are mere sub-offices under larger head post-offices, Ventnor and

HAMPSHIRE
TO LONDON
SOUTHAMPTON
LYMINGTON
HURST CASTLE
Yarmouth
Freshwater
COWES
East Cowes
OSBORNE
RYDE
Pier Head
Sea View
St Helens
Brading
NEWPORT
ISLE OF WIGHT
Sandown
Shanklin
VENTNOR
Niton
EXPLANATION
INSTRUMENTS. Ptg SN ABC
OPEN WIRES.
CABLES.

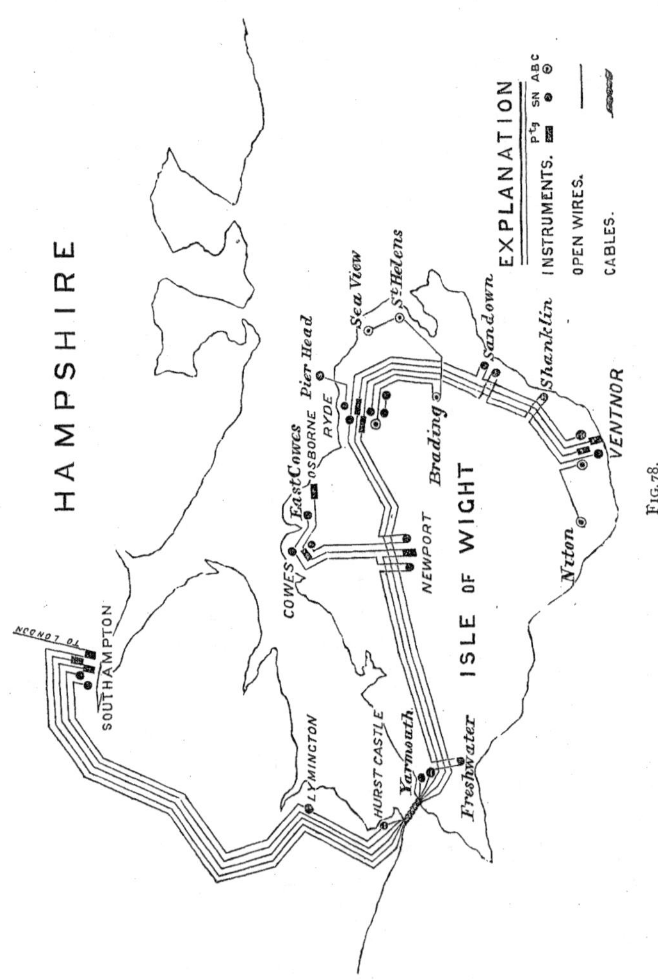

Fig. 78.

Ryde, are amply served by the A B C. East Cowes, West Cowes, Newport, Yarmouth, Freshwater, speak to Southampton by means of a Needle circuit, because the amount of work at the first and the two last offices will not justify the employment of a trained telegraphist. On the contrary, the traffic to and from Osborne, Ventnor, Ryde, Newport, and Cowes, justifies the employment of skilled operators, and they are served by Morse circuits; and the amount of business done at Ventnor and Ryde is such as to require communication with London as well as with Southampton. In some cases, such as Ryde, the business is such that at certain periods of the year it occupies a wire to itself to London, and even this is then insufficient, and additional facilities have to be afforded. A direct wire, with only the terminal offices upon it, and fitted with Sounders, is the most perfect telegraph arrangement we can devise. The insertion of intermediate stations at once reduces its efficiency principally by blocking the wire with local messages. But even a direct wire should be supplemented by a second means of communication in case of failure or accident. Thus each head office like Ryde has a second string to its bow. Telegraphic circuits, even of the simplest and most perfect character, are singularly liable to failure from causes which will be described (§ 297); and occasionally periods of pressure arise from political, special, and local causes, such as elections, races, assizes, &c., and it is imperative that all well-organised systems should be prepared for such emergencies. Thus even the Atlantic Ocean has been spanned by four wires, to secure uninterrupted communication between Europe and America.

116. It is desirable to say a few words upon the proper distribution of battery power upon a circuit; that is, the number of cells which should be allocated to each wire at each station. This depends upon so many conditions of climate, country, size of wire, character of support, &c., that no rule or law can be laid down. In France it has been

attempted to adopt the rule of allowing 10 Marié-Davy cells to every 100 kilometres of line, but this is much departed from. In England the minimum power used is 10 cells of Daniell and 4 cells of Leclanché fitted in troughs, and it usually rises by multiples of this unit. The average is about 5 miles to one Daniell's cell. Hence, if we had two places separated from each other, like London and Liverpool, 210 miles, we should use 40 cells. Every relay is equivalent to about 40 miles of line, and every needle instrument equivalent to 10 miles of line. This must be added to the mileage of wire. Thus between Southampton and East Cowes there are 50 miles of wire and six instruments, equal together to 110 miles of line, requiring therefore 20 cells Daniell or 12 cells Leclanché to work the circuit. This rule is merely empirical. Experience is the best guide to the telegraph engineer of the proper amount of power required. The strength of the signals, and the amount of adjustment required, will tell him at once whether he has used too few or too many cells.

CHAPTER V.

SPECIAL TELEGRAPHY.

A. AUTOMATIC TELEGRAPHY.

117. ALL the different kinds of apparatus which have been described are manipulated by the hand, and though in the A B C and needle systems little skill is needed to work the sending portion of the instruments, yet in the other systems not only skill, but practice and endurance are required to keep up the constant subdivision of time into dots and dashes. The human machine tires, and as a consequence not only is the speed of working reduced, but errors, which

lead to repetitions and delays, are made. The limit of speed with which the hand can work the key of the Morse instrument is soon reached. It is impossible to maintain by hand the maximum useful power of the system. Now signals can be made to follow each other on the simple Morse apparatus far quicker than clerks can send or even write. The muscular motion of the wrist and the directive action of the mind have their limits, both as regards speed and duration. They cannot reach the recording speed of a Morse receiver on a short circuit. But if the manipulation of the human machine be replaced by the precision and regularity of a mechanical machine, not only can we attain, but far exceed, the highest speed of the ordinary Morse or Sounder.

118. Moreover the sending of a clerk after a time loses clearness and legibility, and health, both of mind and body, affects his speed of working. Hence early efforts were made to replace the human machine by some mechanical contrivance which would not only remove the defects inherent to manual labour, but would secure precision in the formation of the characters, accuracy in the despatch of messages, and speed in the transmission of work. Bain was the first to propose this in the year 1846. He punched broad dots and dashes in paper ribbon, which was drawn with uniform velocity over a metal roller and beneath styles or brushes of wire, which thus replaced the key, for whenever a hole occurred a current was sent by the brushes coming in contact with the roller. The recording instrument was his chemical marker (§ 72). The speed at which messages were transmitted at experimental trials was enormous; 400 messages per hour were easily sent; but defects existing in the machinery, which was crude, and disturbances on the line, whose causes were then unknown, interfered with its practical use. It therefore was not persevered with, because really it was not wanted; but now that telegraphic business has increased so enormously that extra wires are needed

in every direction, apparatus which increases the capacity of the wires by sending through them a greater number of messages in a given time has become a necessity of the age.

119. Wheatstone's system of automatic telegraphy is that which is used in England. Bain's method of punching has been considerably modified, and the messages are recorded on an exceedingly delicate form of direct ink-writer.

The apparatus consists of three parts: the *perforator*, which prepares the message by punching holes in a paper ribbon; the *transmitter*, which sends the message under the control of the punched paper when passing through it; and the *receiver*, which receives the message at the distant station when thus sent.

120. *The Perforator*, which is shown in perspective by fig. 79, and in plan and front elevation by figs. 80, and 81,

FIG. 79. $\frac{1}{8}$th real size.

consists of three levers or keys, five punches, and a groove and jerking arrangement to guide and move forward the paper when it has been prepared. The paper *pp'* (fig. 80) is of a stiff white description dipped in olive oil. *a b c* are the three keys which are depressed, and which actuate and drive the punches or perforators through the paper, cutting or punching out clean round holes. 1, 2, 3, 4, 5, (fig. 81) are the punches which perforate these holes in the paper. Key *a* causes 1 2 and 3 to

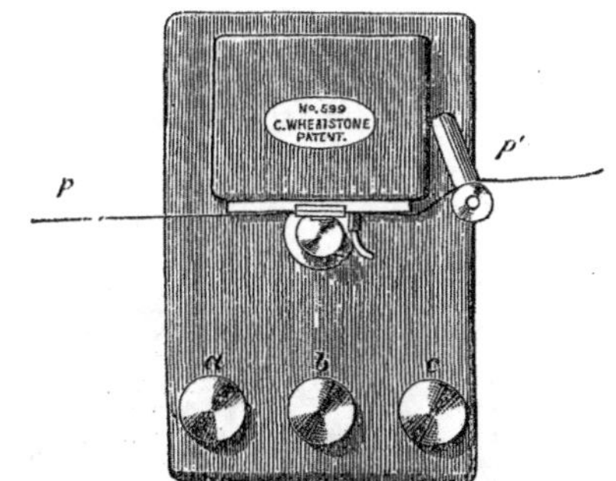

Fig. 80. ¼th real size.

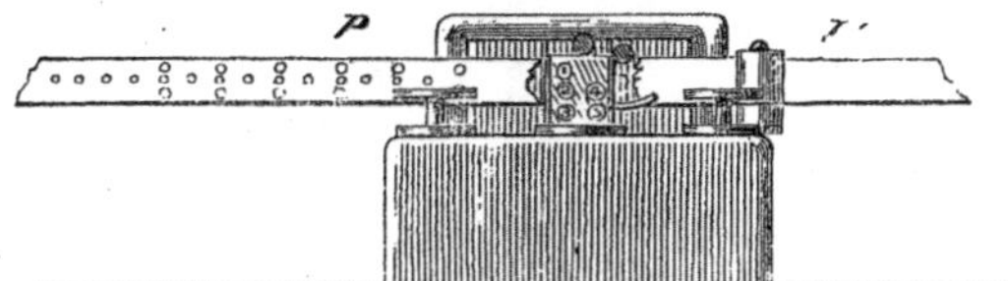

Fig. 81. ¼th real size.

perforate the paper in one vertical line, thus: $\begin{smallmatrix}\circ\\ \circ\\ \circ\end{smallmatrix}$; *b* causes 2 only to punch thus: $\circ$; and *c* causes 1, 2, 4 and 5 to perforate the paper thus: $\begin{smallmatrix} & \circ & \\ \circ & & \circ \\ & \circ & \end{smallmatrix}$; *a* corresponds with dots, *b* with spaces, *c* with dashes. The holes made by *b* are in the centre of the paper, and are smaller than the upper and lower ones made by the other two keys *a* and *c*. They admit the teeth of a little star wheel, which is turned through a small space whenever one of the keys is depressed, and which thus moves the paper on a certain distance for each depression of either key, by a species of rack and pinion movement. The space through which the paper is moved by *c* for a dash is twice the length of that through which it is moved by either of the other keys. In fact two central holes, 2 and 4, are punched for each dash required, and the star wheel is made to turn two teeth instead of one, as in the case of the other two keys. If *a*, *c* and *b* be struck or depressed in succession we have the paper prepared for the letter A; if *c*, *a*, *a*, *a*, and *b* be struck, as indicated by the repetition of the letters, we have the paper prepared for the letter B; and if *c*, *a*, *c*, *a*, and *b* be struck, we have the letter

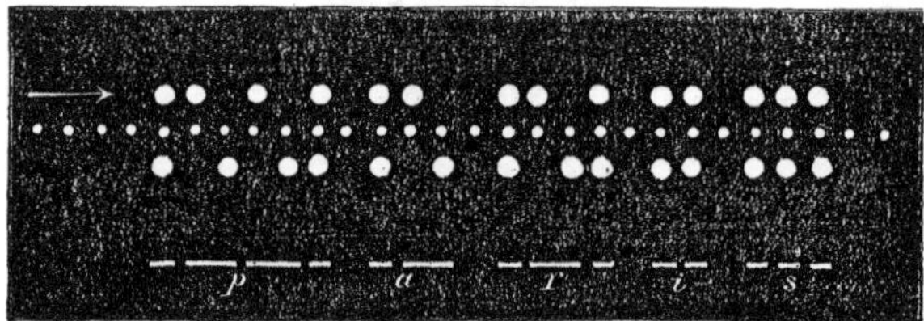

Fig. 82.

C prepared upon the paper. The word *Paris* thus prepared is indicated by fig. 82.

121. It is difficult to indicate these movements by means of a diagram. Their ingenuity, simplicity, and mechanical perfection are best comprehended by an examination of the perforator itself. The keys are usually struck by small mallets grasped by the hands, but at the Central Telegraph Station Mr. Culley has applied the air pressure, employed in the building to work the pneumatic tubes, for the performance of this work. Three piano keys, easily depressed by the fingers, open valves which admit the compressed air into little cylinders fitted with pistons which when forced down depress the keys *a*, *b*, *c* (fig. 80). The labour of punching with the mallets is considerable, and this application of air pressure is very beneficial and is much liked. The power at command is so large that three or even four ribbons are frequently

punched simultaneously at the rate of 40 words per minute. An experienced operator usually punches at the rate of about 45 words per minute on either plan ; but the average rarely exceeds 40.

122. *The Receiver* is a Morse direct inkwriter, of a novel and sensitive character. It is shown in perspective by fig. 83. The paper is drawn above the inking disc with a uniform velocity, and the instrument has a governor which enables the speed of the paper to be regulated so as to allow the instrument to record at any rate of speed between 20 and 130 words per minute. The light inking or marking disc is fixed to an axle geared with the clockwork, and rotates partly within the groove of a larger disc that dips and rotates in the reverse direction in a well of ink. This latter disc takes up the ink by capillary attraction, and thus feeds the marking disc without introducing friction. The axle *a* (figs. 84 and 85) that actuates the inking or marking disc is moved by the armatures T and T′, which are of soft

FIG. 83. ⅛th real size.

iron, and which are maintained in a magnetized condition by the permanent magnet N S. The mode of adjustment is also shown in the diagram. The armatures are so arranged that when once a current, however short in duration, has moved them, by passing through the electro-magnet E, they remain as placed. They can only be restored to their normal position by a current in the reverse direction to that which sent them over. Thus when once the armatures and the axle have been moved, the inking disc has been brought in contact with the paper, and a line is recorded upon it, until a reverse current is sent. A dot is made by sending a short momentary current in the proper direction to move the disc, and instantly afterwards another momentary current in the reverse direction to bring it back again. A dash is made by sending this reverse current after the first current at a longer interval. Thus two short momentary currents are required for every mark made ; and these currents are in the opposite direction, and separated from each other by varying intervals of time. It is the punched paper that determines the intervals that separate these currents, and the transmitter is the instrument that sends them.

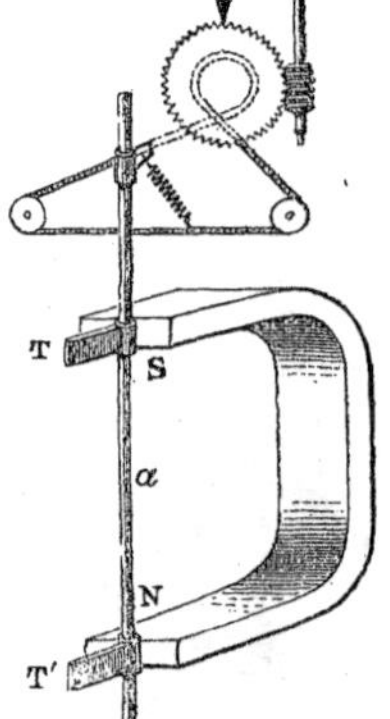

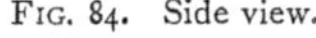

FIG. 84. Side view.

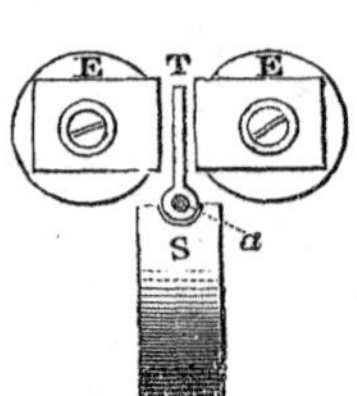

FIG. 85. Top view.

123. *The Transmitter* replaces the key of the ordinary apparatus, and it sends the currents under the control of the punched paper by mechanical means. Hence the name of the system—*the automatic.* The currents that are transmitted are of different duration to those sent in any of the

systems which have been described, and the apparatus for sending them is complicated. The principle of the portion of the apparatus sending these reverse currents is shown in figs. 86 and 87. Let D be a metal disc divided into two portions, which are insulated from each other, and which have metallic pins, *a* and *b*, fitted to them. Let the portion marked *b* be in connection with the line, and that marked *a* with the earth. Let the ends of two rocking levers, by the tension of the springs *s* and *s*′, rest on these pins, one lever C being in connection with the copper pole of the battery B, and the other lever Z with the zinc pole.

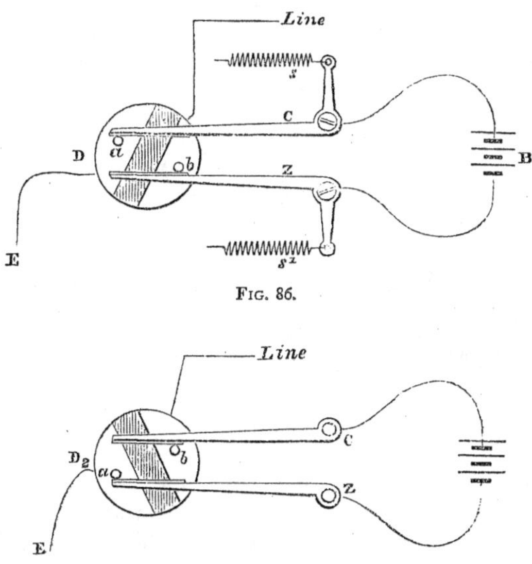

Fig. 86.

Fig. 87.

Now it is evident that in the first case illustrated (fig. 86) a negative current is going to the line, and in the second case (fig. 87) a positive one. If, therefore, the disc D be made to vibrate between the position shown in D (fig. 86) and that shown in D_2 (fig. 87) regularly and continuously, a succession of currents in reverse directions will pass to the line; and if a Wheatstone receiver be fixed at the distant end of the line a succession of dots will be recorded upon it. If, however, before the disc reaches the position D_2 the line or earth wire be temporarily disconnected, a reversal will be missed and a *dash* will be recorded at the distant station instead of *two dots*. The function of the punched paper is so to regulate the motion of the transmitter as to produce this disconnection at the proper moments, and thus cause the currents to flow in such a way as to form dots and dashes.

124. The punched paper PP (fig. 88) is carried forward, in the direction shown by the arrow, by a similar little star wheel to that which moves it in the perforator (§ 120), by catching

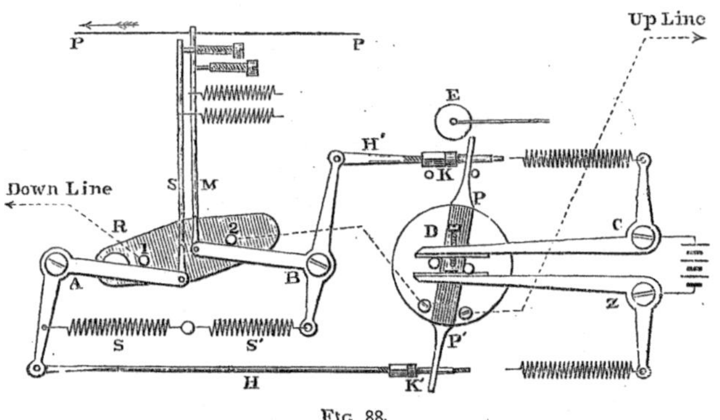

FIG. 88.

hold of the central row of holes. Two rods, M and S, are fixed to the ends of the rocking levers, A and B, which are attached loosely to the framework of the instrument, and are maintained at a constant upward pressure by means of the spiral springs S and S'. The two rods M and S play one opposite each of the two lines of larger holes in the punched paper, so that, unless otherwise restrained and regulated, their ends would alternately project through them if there were holes, or would be checked by the paper if there were no holes. The rod M is shown projecting through the upper hole, and the adjustable studs and springs keeping them in position are illustrated. R is a beam of ebonite which has two metal pins, 1 and 2, fixed into it, and which is maintained in a constant condition of equable vibration by means of the moving train of wheelwork of the apparatus. These pins rest upon the crank levers A and B, which are by this means kept rocking in unison with R. The lever A has at its lower end a rod H fixed to it, and the lever B has a similar rod H′; the ends of these rods pass loosely through the pieces P and P′ fixed to the divided disc D; the collets K, K′ by pushing P and P′, cause that disc also to vibrate in unison with R. This disc, with the levers C and Z, is identical with that illustrated in fig. 86. The lever C has a screw fixed into it, which rests, when the disc is in the middle of its vibration, upon a little piece of ebonite fixed on Z, to prevent the battery being short circuited. The roller E completes the task commenced by the collets K and K′, and retains the disc in its position until it is reversed by the thrust of the collets K or K′. Thus this beam R is constantly and regularly causing the rods S and M to be moving up and down, the levers A and B to be rocking to and fro, and, so long as the play of the rods S and M is unrestrained, the disc D to be vibrating backwards and forwards, moving with it the levers C and Z.

125. The pin 1 on the beam R is in electrical communication with the down line or earth, and the pin 2 with one half the disc D, whose other half is connected with the up line. When there is no paper in, the pins are in constant and unbroken contact with the levers A and B, which are carefully platinized, and which are connected together by the framework of the instrument and by the springs S and S′, thus maintaining the circuit complete and sending a succession of reversals producing dots at the distant end. When there is paper in, however, and this paper is not punched, it limits the upward play of the rods, the levers A and B do

not follow the pins throughout their full phase, the disc D remains unmoved, connection is destroyed, and no currents go to line. If the paper be punched, then a dot $\overset{\circ}{\circ}$ allows a reversal to be made at each vibration, and a dash $^{\circ}\overset{\circ}{\circ}$ causes it to be made at each second vibration. The pin M enters the upper hole, and the marking current is sent, the pin S enters the lower hole, and the reversing current is sent.

126. The transmitter is regulated to transmit at a maximum speed of 130 words per minute, but this by no means indicates the absolute speed at which the instrument can work. Its maximum rate of sending is however determined more by the rate at which the receiver can receive, and this is limited not only by the mechanical inertia of the instrument itself, but by what may be called the magnetic inertia of its electromagnet. An electromagnet, when connected up in a telegraphic circuit, cannot be magnetized and demagnetized with infinite rapidity. The core takes time to magnetize and to lose its magnetism, and the currents which pass through the coils generate or induce other currents in the same coils, which have the effect of retarding this demagnetization. If the transmitter were to work at a higher rate than 130 words per minute, the receiver would simply register a long line or no line at all, because sufficient time was not allowed for the reverse currents to produce their effect on the armatures that actuate the inking disc. Thus the rate at which the transmitter works in practice is limited by the rate at which the receiver can register its signals.

127. Such an apparatus can send uninterruptedly messages at the rate of 130 words per minute as long as it is working on a short line ; but as we increase the length of our line, so we diminish the rate at which it works from causes which have been already referred to in § 108. The currents to produce the motion of the inking disc at the distant station must be of longer duration on longer lines to over-

come the disturbance due to induction, and hence the vibrations of the rocking-beam must be slower for long lines than for short ones.

Thus while a speed of 120 words per minute can be obtained between London and Manchester, we can only obtain 90 between London and Sunderland, 60 between London and Aberdeen, and between London and Dublin only 40 words per minute.

128. When a quantity of electricity flows through a line in the form of a current, the first portion of the current is retained or *accumulated* upon the surface of the wire, in the same way that a charge is retained or accumulated upon the surface of a Leyden jar. The quantity accumulated depends upon the length and surface of the wire, upon its proximity to the earth, and upon the insulating medium that separates it from the earth. Thus, in the case of a submarine cable, the conductor of which is insulated with *guttapercha* or *indiarubber*, and is maintained in very close proximity to the earth, a very considerable charge is held by the wire. An overground wire is insulated in air, and though it is maintained at a considerable distance from the earth, yet it is in close proximity to other wires which are in connection with the earth, and it also retains a charge. In fact it is found, in England, that the charge retained by twenty-three miles of ordinary line wire is about equal to that retained by one mile of a cable of average dimensions. This power of retaining a charge is called *the electrostatic capacity* of the circuit.

129. Now what are the effects of this electrostatic capacity? In the first place, as we have already pointed out (§ 108), it absorbs all the electricity of a short momentary current and prevents the appearance of any current at the distant station. And as it absorbs the first portion of every current sent, it has the same effect as if it retarded or delayed the first appearance of the current at the distant end. Thus the apparent velocity of the current is diminished

by the amount of induction present in the circuit. In a circuit of very low capacity there is practically no induction, and the current appears instantaneously at the distant end. In a circuit where there is capacity there is induction, and the first appearance of the current is retarded according to the amount of induction present. Thus between Europe and America, on the Atlantic cable, the current is retarded four-tenths of a second.

130. In the second place, before a current in the reverse direction can be sent through the circuit, the whole of this charge upon the wire must be withdrawn or neutralized before a second charge of opposite sign can be accumulated upon it. This discharge occurs in two ways, either as a current flowing out at each end to earth, or by the neutralization of the reverse current by one of opposite sign. In the first case the current flowing back to the sending station is called the *return current*, and that flowing out at the receiving station continues or prolongs the primary current. In the second case the first portion of the reverse current is occupied in neutralizing a portion of the opposite charge, as well as supplying that required by the capacity of the circuit, while the other portion flows out at the distant end, producing as before the *prolongation* of the current. If in the first case one end of the wire, say that at A, be disconnected, all the charge flows out at the distant end, and the prolongation of the current is increased. Thus the second effect of electrostatic capacity is to prolong the primary current at the distant end. Charge therefore produces *retardation*, and discharge *prolongation*.

131. These effects of induction are beautifully shown and can be clearly studied by means of Bain's chemical marking apparatus (§ 72). Fig. 89 gives a representation of the effect which would be observed with dots (1) on lines of little induction; (2) on lines of moderate induction—say 300 miles of overground wire; and (3) on long cables. While 20 dots per second can be firmly and clearly recorded on the

first line, 10 are recorded on the second line, and only 2 on the third. If more than 2 dots per second in the last case or more than 10 dots per second in the second case be trans-

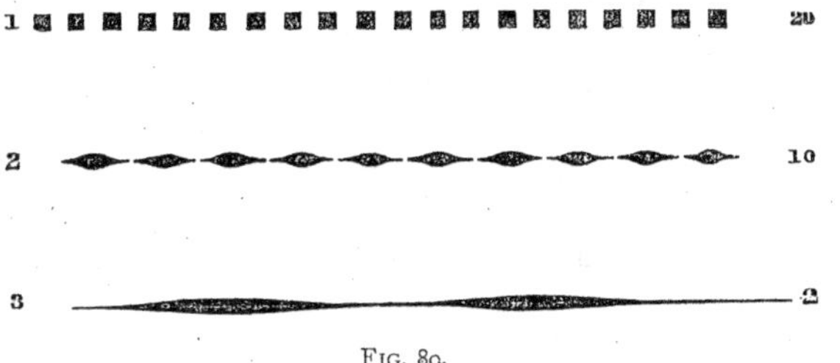

Fig. 89.

mitted, then the marks run together, they become illegible, and on a Wheatstone receiver they form a continuous line.

132. If dashes be considered in place of dots the effect on speed of working is still more evident. With dots the current at the sending end can be regulated in duration, so as to allow just sufficient current to appear at the distant end to record its arrival, and no more. The line is only partially charged; but with a dash, the current being longer in flowing, the line is more fully charged, and therefore a greater quantity of electricity has to be neutralized or withdrawn before a second current can flow, and consequently more time is consumed in sending. Hence any system involving dashes directly sent is objectionable on lines containing submarine cables. Such lines should invariably be worked with currents of uniform duration. This is one merit of the Wheatstone transmitter. The dots and dashes are produced by currents of equal duration, but they are separated from each other by periods of unequal duration, and thus the condition of the line at the commencement of each current is variable. If a succession of dots be uniformly sent into a neutral line at the maximum speed which the receiver at the distant end can register, it by no means follows that between each dot the line must be completely discharged; it is only necessary

to reduce the potential so as to produce that difference between the ends of the receiving apparatus which will reduce the current low enough to allow the antagonistic force, whatever it may be, to act upon the armatures of the relay.

133. Again, the inductive capacity of a line is unequally distributed. Its rate of working is naturally affected by this distribution. A circuit may be made up of overground wires, underground wires, and cables. Since cables have the largest capacities, it is their position which materially influences the speed of working. Between London and Amsterdam there are 130 miles of land wire over the Great Eastern Railway, then a cable 120 miles long, and then 20 more miles of land wire. London could send to Amsterdam only 30 words per minute, while Amsterdam could send to London 45. Between London and Dublin there are 266 miles of land wire in England, 66 miles of cable and 10 miles of land wire in Ireland. Dublin can send to London 80 words per minute, while London can send to Dublin only 40 words per minute. This difference is due principally to the influence which the long overground line has by its leakage in favouring discharge direct to earth without influencing the receiving apparatus.

134. These effects of induction influence the working of automatic circuits in various ways. In the first place, the retardation and prolongation of the currents diminish the rate of working; fewer currents can pass through in the same time. In the next place, they distort the appearance of the marks made by the currents at the distant end, either by the loss of dots, or by the running together of dots and dashes, or by the conversion of dashes into dots and of dots into dashes. Letters are thus deformed and even converted into other letters. A dot entering a neutral line becomes a dash from prolongation; a dot following a dash may be lost because its current is entirely occupied in neutralizing the return charge of the dash, or it may be only clipped as it were, which may also be the case of a dash following a dot. Dots and dashes

following each other too rapidly run together, because there is no time for discharge and reversal.

135. How are these defects in working provided against? They cannot be entirely remedied, because they are inherent to the principle of working, but they can be counteracted, and their ill effects can be reduced, by an exceedingly simple contrivance, due to Mr. Culley. Between the two pins of the rocking lever R is inserted a set of resistance coils *R*, shown in fig. 90, which can be arranged so as to introduce no resistance at all, or resistance varying up to 10000$^{\omega}$ or infinity. When there is no resistance inserted the transmitter is converted into a reversing key (§ 108), for there is no disconnection of the line circuit when the rocking levers A or B fail to follow the rocking beam R. As resistance is inserted so the current is varied from an initial current of the normal strength, followed by one in the same direction of a weaker strength, which in its turn is succeeded by a reversing current of normal strength, also followed by one in the same direction of weaker strength and so on. Every circuit requires a different current of different strength to produce this compensation, and this varies with weather and other conditions. This change in the strength of the compensating current is easily effected by the variation of the resistance coils, and a few trials soon show the proper arrangement to produce the maximum result. On an average about 6000$^{\omega}$ are inserted in *R*; these weak compensating currents maintain the potential of the line at an equable height, so as to cause it to fall uniformly and to be nearly in the same condition at the commencement of each signal.

136. Fig. 90 gives a plan of the way in which a complete set of the automatic apparatus is fitted up at one station. It contains a plan of the transmitter switch, which is shown in the position it acquires when the starting handle is turned to allow the machinery to run and the currents to flow. It also shows a double current key, so connected up as to be thrown into circuit by the above switch for speaking

UP LINE OR EARTH
R
PLAN OF TRANSMITTER SWITCH.
BATTERY
KEY
GALVR
DOWN LINE OR EARTH
SOUNDER
LOCAL
RECEIVER
S
C
Z
E
K
K'
P
P'
D
H'
H
B
A
R
S
M

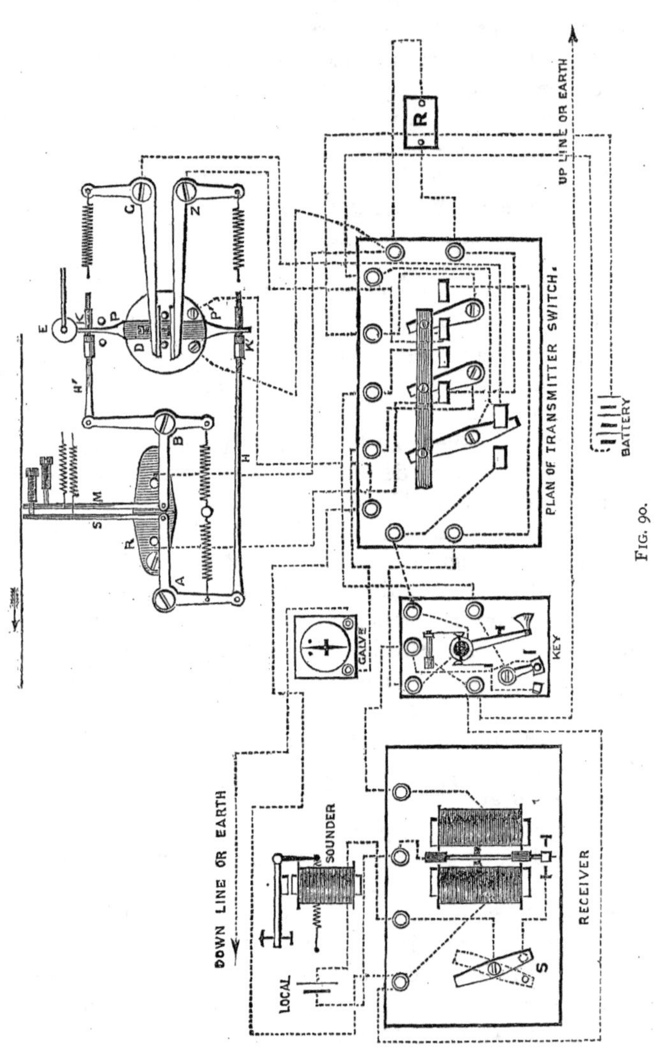

Fig. 90.

purposes when the transmitter is not running. There is also a galvanometer to show that currents are properly going and coming.

137. The receiver has also a switch S in connection with the starting handle, which when it stops throws a sounder into a local circuit, used not only as a call or alarm, but as a receiver when the key is used, or when the line circuit is used for ordinary working. When the receiver is running the switch throws the sounder out of circuit.

The local circuit is completed by the electromagnet of the receiver, which thus acts as a relay as well as a direct inkwriter.

The transmitter is shown in the neutral position when no current flows to line, and when the pin on the lever C is acting to prevent the battery from being placed on short circuit.

R shows the resistance coils employed for producing the compensation currents.

138. Automatic instruments are used on nearly all long circuits in England, not only because they increase the capacity of the wires for the conveyance of messages, but because they are so specially adapted for the conveyance of news, which is such a distinctive feature of the English system of telegraphy. One strip of punched ribbon will do for any number of circuits, and thus the labour of preparing the slips for transmission is very much reduced. In fact, without this system it would be simply impossible to transact the enormous amount of intelligence sent through the wires all over the country. There are many news circuits radiating from the Central Telegraph Station, having three and four intermediate stations upon them, one or more of which repeat or translate onward to three or four more stations. Thus one punched slip disseminates the news to many places.

139. It is of course evident that the chief value of the automatic. system, beyond its extreme accuracy, is its increased speed of working. It may be said to double the

capacity of wires. The average rate of automatic working in England due to the length of circuits and to the amount of inductive capacity present is about seventy words per minute. Thus one wire fitted with the automatic apparatus can do the work of two fitted with the ordinary apparatus. But the former involves additional expense in working and additional delay to each individual message. When a wire is kept going at its full speed two punchers, one adjuster or sender, and three writers, are employed. Five additional clerks are therefore required at each terminal station. The messages are punched and transmitted in batches of five or six. Thus a message has to wait to be punched, and to take its turn in its batch. This involves delay. For these reasons it is not economical to introduce automatic working on short circuits, except for special occasions and for breaks down, and hence it has been confined principally to long circuits.

140. The automatic system has proved invaluable when a sudden glut of work has been handed in at a station, or when the communication has been interrupted through storms and accidents. Once, when four out of the five wires between London and Birmingham were broken down, the remaining wire, working automatically, did the work of all, but of course with some delay.

Although the system is automatic, we are not entirely free from human labour or from the frailty of human nature. Whenever we are subject to the action of the will we are liable to error. Thus dots are sometimes punched for spaces, and vice versâ, and absurd errors are made.

B. Submarine Telegraphy.

141. Submarine cables of considerable length like those connecting Europe and America, or those forming the great chain connecting the Mother Country with the Antipodes, have to be worked on a peculiar method, specially devised to obtain out of them the maximum speed of working. Re-

lays or other forms of apparatus, whose action is dependent upon electromagnetism, are inadmissible for various reasons : 1st. They require currents to influence them stronger than can with safety be transmitted through long submarine cables. 2nd. They aggravate the effects of retardation. The causes of retardation in such cases have been sufficiently dwelt upon (§ 128–130), but there are other causes of embarrassment which have also to be provided against. Different portions of the earth, from causes which are not yet known, are frequently at different potentials. When these portions at different potentials are connected together by wire, we have what are called *earth currents.* The currents vary in strength and duration during different periods of the day and year, and at certain seasons they acquire such magnitude as to be called 'electric storms.' They then interrupt the circuits to such an extent as to render working difficult and even impracticable. On long cables they are specially prevalent, and become of such strength as to endanger the safety of the cable. They are to be guarded against in two ways : 1st, by dispensing with the earth and using a second wire as the return wire, working, as it is called, in a *metallic circuit.* 2nd, by using *condensers* and working with a broken or interrupted circuit.

142. The first method is used chiefly on land lines because it can be easily and rapidly resorted to, and on cables when there are wires available ; but the second method is that which is principally used on cables, and it is very effective. It was invented by Mr. C. F. Varley.

A *condenser* or *accumulator* is a term applied to an apparatus composed of alternate layers of tin-foil and paraffined paper (or mica), so arranged as to form a flat Leyden jar of large surface, and constructed to give us whatever capacity we require. $a\ a_1\ a_2$, $b\ b_1\ b_2$, (fig. 91), are square pieces of tin-foil, separated by sheets of thin paper steeped in liquid paraffin or some other preparation. The series $a\ a_1\ a_2$ are connected together, and so are the series $b\ b_1\ b_2$. A and B thus become connected with what may be called the inside and outside coatings of a Leyden jar; and by putting B to earth, or to the other pole of a battery, we can communicate

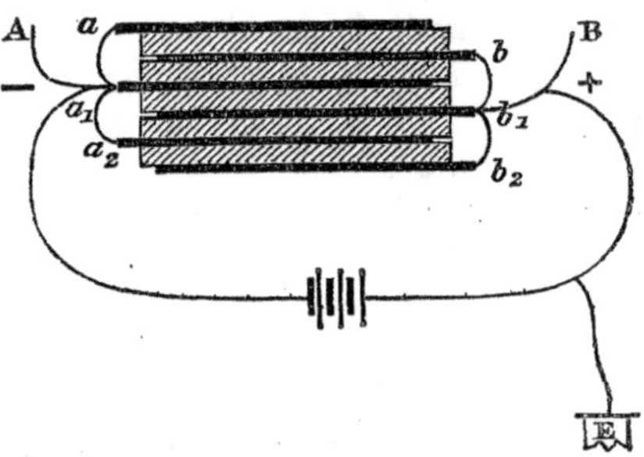

FIG. 91.

to A a charge, the quantity of which will depend upon the number of cells used, upon the surface of the plates opposed to each other, and upon the number of the series forming A and B. Thus we can construct condensers of any capacity, giving a charge varying from that accumulated upon one mile of overground wire up to that accumulated upon an Atlantic cable. The unit or standard of reference by which capacity is known is called the *microfarad*, and it is equivalent to the charge contained by about three miles of cable.

143. Condensers are conventionally represented by *c*, fig. 92. Let A B be a wire connecting Europe and America, K an ordinary key, and B a battery at A, C a condenser inserted in that wire, and G a galvanometer at B. Now if the circuit be as arranged, it is evident that it is broken at C, no continuous current can pass from A to B, and thus earth or other extraneous currents are prevented from flowing through the galvanometer. But how can we affect the galvanometer G at B? In this way: when we depress the key K, a current flows into the cable to charge it; one side *a* of the condenser is thus connected with one pole of the battery, its potential is raised, it is charged say negatively. The ne-

gative charge accumulated on *a* attracts across the paraffin a positive charge on *b*, and repels a negative charge. This

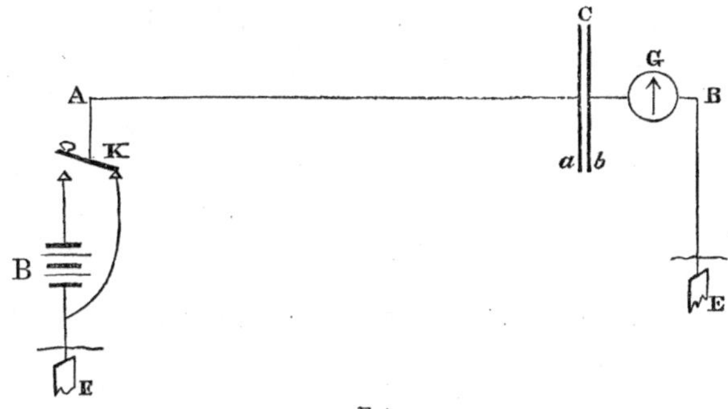

FIG. 92.

positive charge apparently passes from earth at B through the galvanometer in the form of a short current or pulsation. When K is released and falls back to its normal position, the cable is discharged, the potential of *a* is again reduced. The positive charge on *b* is released, and it flows to earth in the reverse direction through G in the form of a second current or pulsation. Thus whenever we depress the key we affect the galvanometer with a reversal.

144. The condenser might be placed at the sending end,

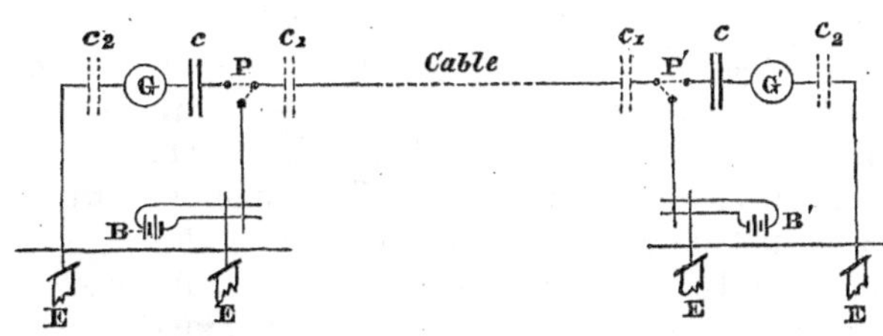

FIG. 93.

but it is better to employ condensers at each end, as shown in fig. 93. The arrangement for working a submarine cable by means of condensers is there symbolically represented. The condenser may occupy any one of the positions marked c c_1 c_2. When in the position c_1 the sending current proceeds from the discharge of the condenser, and works the apparatus at the distant station, the cable there being disconnected by means of the switch P from the commutator and placed in connection with the receiving galvanometer G. If the condenser occupies the position c, the galvanometer is worked direct by means of the discharge from it: if it occupies the position c_2, the charge into it works the galvanometer.

145. Now by using galvanometers or other receiving apparatus of the most sensitive character, which will be actuated by the first appearance of the current, we are able to work cables with the smallest possible electro-motive force. This not only conduces to the safety of the cable, but adds to the speed of working.

Thus by suitably determining the size of the condenser, the number of cells, and the delicacy of the galvanometer, we can transmit signals which shall give us the maximum speed with the minimum expenditure of power, and thus effectually counteract the ill effects of earth currents, and reduce to the lowest possible point the retarding influence of induction.

The condensers used have a capacity of 20 microfarads, which is equivalent to the capacity of about 60 knots of cable, and from four to ten cells of one or other of the form of Daniell's battery—generally Minotto's (§ 42)—are employed.

146. The galvanometer that is used is Thomson's reflecting galvanometer—the most delicate and perfect instrument of its kind ever invented—without which long cables could scarcely have been made commercially successful. The needle is simply a piece of watch-spring $\frac{3}{8}$ inch in length, suspended by a short thread of cocoon silk without torsion, and having a circular convex mirror of silvered glass cemented to it. It weighs only $1\frac{1}{2}$ grains. It is

suspended in the centre of a coil of very fine wire, giving a resistance of about 2000$^{\omega}$. A magnet, which is raised or lowered above it, or turned around its axis by means of a screw, exerts a directive force on the needle, so as to cause it to reflect the thin image of a lamp passed through a fine slit on the centre of a scale. It also controls the vibrations of the needle so as to make its movements almost a dead beat; indeed they are sometimes so sudden and short as only to broaden the spot of light.

This directive magnet, instead of being fixed over the coil, is sometimes placed independent of it by the side of the instrument. It is thus shown (A) in fig. 94, which represents the arrangement of the apparatus at one end of a long submarine cable. G is the galvanometer, one end of which is attached to the condenser C, and the other to earth, by means of the earth terminal T. B is the battery which is connected to M, the transmitting portion of the apparatus, and similar in every respect to the pedals of the single needle instrument described in § 48. R is the resistance coil employed in connection with the condenser C, and inserted for adjusting purposes, to suit the varying strength of the currents and conditions of the cable. P is a small switch to which the cable is brought, and by means of which it may be put direct to earth, or placed in connection with the galvanometer G, or the keys M, according as it desired to receive or to send.

The beam of light proceeding from the lamp D, through the slit *f*, is concentrated, by means of a lens L, on to the mirror *m*, whence it is reflected back to the scale E. By means of the movements of this reflected beam of light to the right or left the alphabet is formed in precisely the same way as by the motion of the pointer on the dial of the single needle. H is a large box which acts as a species of darkened chamber, and enables the movements of the spot of light to be discerned with ease.

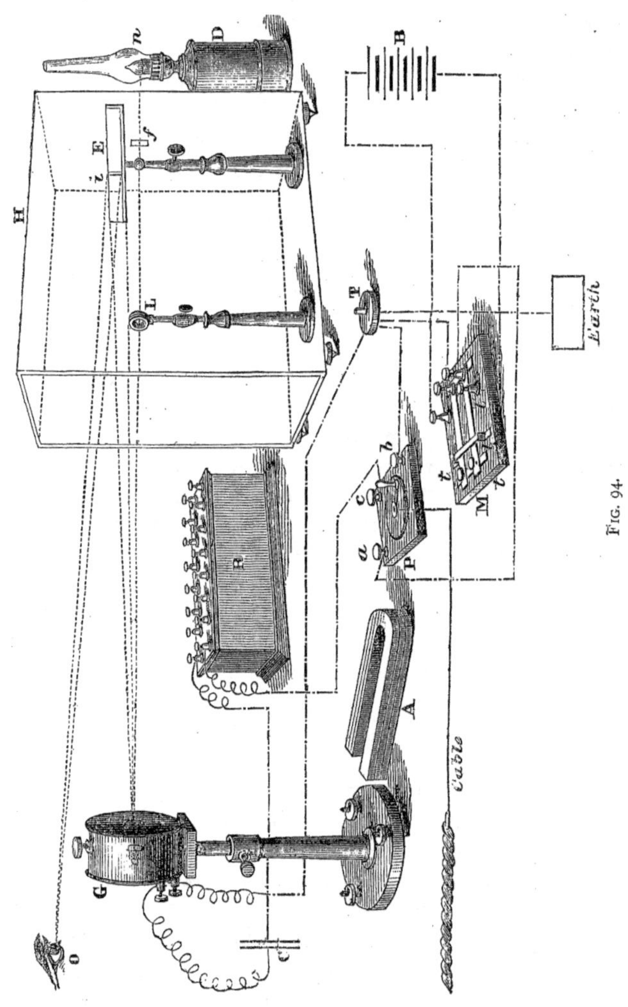

Fig. 94.

A glance at fig. 94 will serve to show the electrical connections which are required. The cable is brought to the switch P, and if it is desirable to put the line direct to earth for either testing or protective purposes the switch bar is carried to *b*, which by means of T is in connection with the earth. When signals are to be received the switch-bar is placed in connection with *c*, and in this way the cable is placed in connection with one side of the condenser through the resistance coils R : the other side of the condenser being in connection with earth through the galvanometer G. If, again, signals are to be sent, the switch-bar is carried to *a*; to this same terminal one of the bars of the manipulator M is connected, and in this way the signals are sent direct to the cable without influencing the galvanometer G.

147. If the ordinary apparatus used for land telegraphy, such as the Morse or Sounder, were used on the Atlantic cables, a word a minute could scarcely be obtained; with the mirror instrument fifteen words are easily sent in the same time, and twenty-four have been obtained. The mirror is really a single-needle instrument, whose index is a spot of light; but apart from its excessive delicacy, it has this advantage over the vertical needle, that in place of having a fixed *zero* or *neutral* line, to the right or left of which the needle vibrates to impart its signals, the zero line —when condensers are not used—moves with the spot of light and wanders all over the scale, the signals being made by the pulsations or vibrations of the spot, and being read by their direction and not by their position or amplitude. Thus signals need not be read by separate distinct currents, as in land lines or when condensers are used, but by the increment or decrement of one continuous current which is continuously flowing out of the cable from its great capacity, and whose potential only is varied by the reversals made at the sending end.

147. The use of condensers, as shown in fig. 93, practically fixes the zero line of the mirror, for it is evident that there will

be no indications of a continuous current now, and that the condenser will only respond to the changes of potential of the current in pulsations corresponding to those imparted to *c*. Sir William Thomson has introduced an instrument which records these signals by spurting ink by electrification, upon a moving paper ribbon, from a fine glass tube which is moved to the right and left by these reversals. A sentence sent by this syphon recorder, as it is called, is shown in fig. 95.

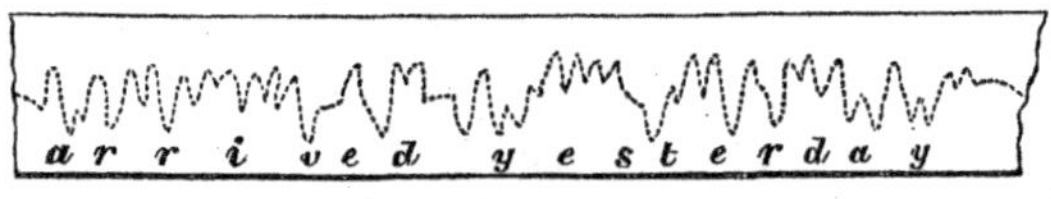

FIG. 95.

C. DUPLEX TELEGRAPHY.

149. The rapid increase in the business of telegraphy has called forth the exercise of the ingenuity of telegraph engineers to increase the capacity of a single wire for the transmission of messages. The automatic system effects this object in one way, duplex telegraphy effects it in another. The first practically enables twice as many messages to be transmitted in the same direction in a given time, the second enables an equal number of messages to be sent in the opposite direction *at the same time.* Both systems have thus doubled the capacity of wires.

150. The duplex system means the transmission *on the same wire* of a message from station A to station B, while B is sending another message to A. Under ordinary circumstances, when A is working to B on the open circuit principle (fig. 70, § 101), any interference on the part of B destroys A's current, and prevents marks from being properly recorded. B's current neutralizes A's, and vice versâ, A's current neutralizes B's ; the result being disturbance and loss of marks. But if we are able to register marks in a similar manner by this neutralization of the current as well as by its formation,

duplex telegraphy becomes possible. There are several modes of doing this, but we shall confine ourselves to a description of those most generally used in England. There are two methods in practical use, the one based on the principle *of the Differential Galvanometer*, the other on the principle of the *Wheatstone Bridge.*

1. *The Differential Principle.*

151. If two circuits of precisely equal resistance be open to a current it will divide itself equally between the two, and the currents in each wire will be exactly equal. If, for instance, the wire Z *l* E, fig. 96, offers the same resistance as the wire Z *r* E the current in *l* will have precisely the same strength as the current in *r*. Now let the electromagnet E, fig. 97, be similarly wound with two wires of equal length, one of which is in connection with *l*, and the other in connection with *r*. If the current through *l* traverse the electromagnet in the *reverse* direction to that through *r*, it is evident that as the currents are equal the polarity induced by the one current must be exactly neutralized by that induced by the other current, for the effects are equal and opposite, and there is no magnetism excited. Thus, as long as the two circuits are intact the currents which flow do not affect the electromagnet; but if the currents in *r* are interrupted, those in *l* will excite the electromagnet, and if those in *l* are interrupted, the currents in *r* will excite the electromagnet.

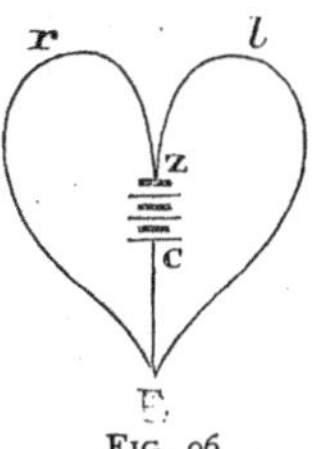

FIG. 96

152. Let A and B, fig. 97, be two stations connected together by the line wire *l*. Let E be an electromagnet at A, such as that just described, and E′ a similar one at B, K a key, and B a battery. Let *r* and *r′* represent resistance coils or artificial lines, each giving a resistance equal to the line circuit. Now let us in the first place assume A alone to be

working to B; every time the key K at A is depressed a current is sent from A's battery. This current divides at

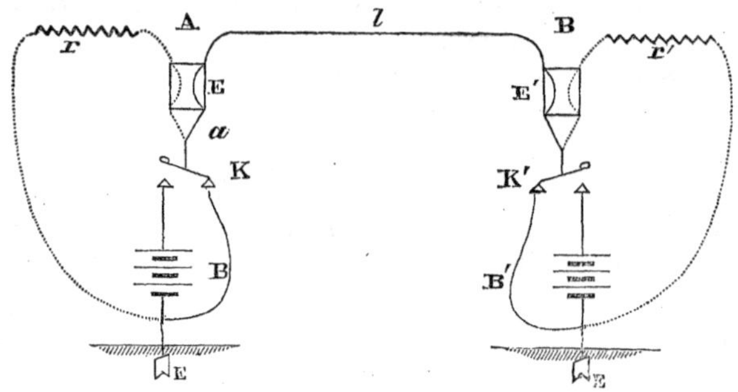

FIG. 97

a, the one half going through the wire in connection with *l* in E, through *l*, and at B, through the wire in connection with *l* in E′, through the key K′ at B to earth and thence back to the battery. This is called the *line current*. The other half, which is called the *compensation current*, passes around the electromagnet E through the wire in connection with *r*, through *r* and back to the battery. As these two currents are equal their effect on E is *nil*, but the line current passing through one coil only of E′ operates it and causes signals to be given. Thus while A telegraphs to B its own instrument is not affected, but that at B is actuated. Similarly, when B alone is working to A its own instrument is not affected, but that at A is actuated. But when B is working to A at the same time that A is working to B, what happens? Every line current that leaves A at the same time that a line current leaves B is neutralized. The compensation current at A is now able to excite the electromagnet, and the armature is moved *in precisely the same way as if* B's *currents were received.* In the same way B's line currents are neutralized, and its compensation currents move the armature of E′ in precisely the same way as if A's currents were received. Thus E and E′ continue to be worked by their respective stations, regardless of the facts that the line currents are being continually neutralized, that practically no current flows between A and B, and that they are operated sometimes by the line current and sometimes by the compensation current. Thus, while A sends messages to B, B can be sending messages to A upon the same wire and at the same time.

153. We assumed that the line current received at A from B was exactly equal to that proceeding from A to B, and that therefore they were exactly neutralized, but it is not so in practice, for owing to the effects of bad insulation the incoming line current is always weaker than the outgoing one. Hence the current received at A from B does not neutralize the whole of the current from A to B, but only a portion of it. It therefore weakens it so that the compensation current preponderates over this resultant current, and registers its signals by this preponderance: The *difference* in the strength of these two currents when both stations are working is very nearly equal to the strength of the current received at A when B alone works, so that the marks, whether made by the received line current or by the preponderating compensation current, are practically the same.

154. We have shown in the diagram that the same poles of the battery are to line, and that therefore the line currents flow in opposite directions; but the same effects occur if the opposite poles are to line, and the currents flow in the same direction. If the current from B flows in the same direction as that from A, the effect, when the two stations work simultaneously, is not to weaken the resultant current, but to strengthen it, and therefore to produce a preponderance of the current in wire *l* over that in wire *r* of relay E, and consequently to register signals; but in this case the marks made at A when both stations are working simultaneously, are not made by the preponderance of the compensation current over the line current, but by the excess of the resultant line current over the compensation current.

155. We have shown the keys K and K′ as putting the line to earth through the back contact 2, but there is an interval when the keys are being depressed when this connection is broken. In fact there are three positions which the key takes up during the operation of sending, viz., 1st when resting against the back contact; 2nd when resting upon the front contact; 3rd when disconnected from both contacts. The line current is not, however, in either case interrupted. The first and second cases are clear, but consider the third; take the key K (fig. 98) and depress it; then the received current, when it arrives at 3, instead of going to earth through 2, returns through the electromagnet by means of the compensation circuit, thence through the resistance R to earth. This continues the effect of the line current upon the electromagnet; for though the resistance to the line current is twice as great, and the current itself is reduced one half, it passes through *both* wires of the relay in the proper direction to actuate the armature, and therefore as the effect is doubled it leaves its influence unimpaired. In fact, it is possible to dispense with the back contact 2 altogether; and this is sometimes done for short lines, but for long lines it is objectionable, for it introduces elements of irregularity which tend to diminish accuracy and speed of working. For accurate duplex working the total resistance of the circuits should be disturbed as little as possible, hence the portion of the circuit from the back contact of the key to the earth should be made equal in resistance to that of the battery R. The best form of key is that which does not

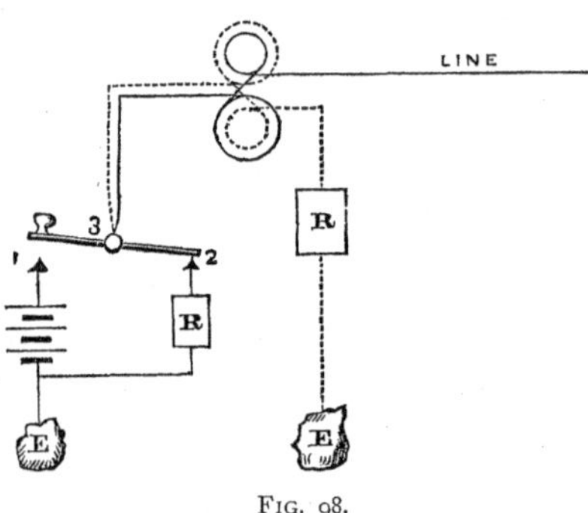

FIG. 98.

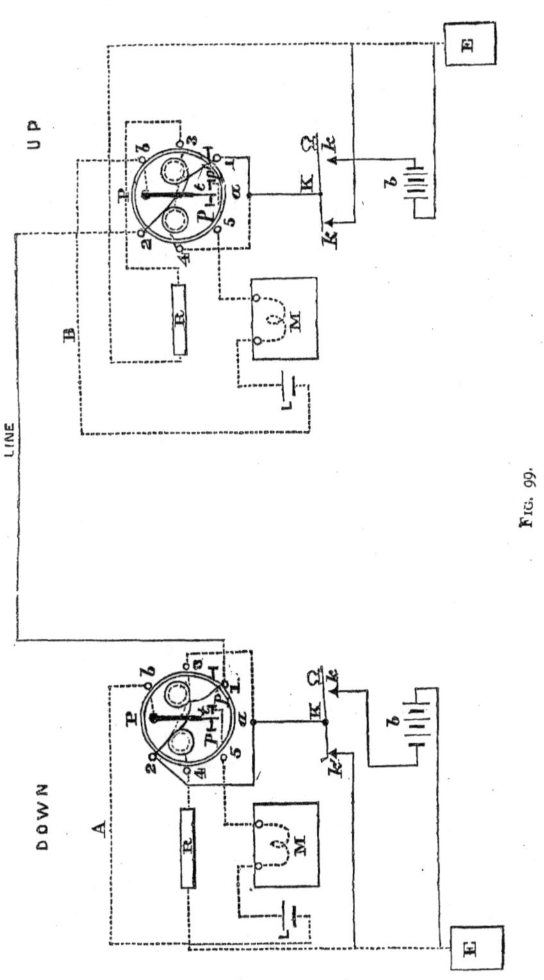

Fig. 99.

break the back contact until the front contact has been made (fig. 103).

156. This explains the principle of the system, but it does not illustrate the apparatus in action. E may be the electromagnet of a direct inkwriter, or it may be a relay. As most circuits are worked with relays we will illustrate such a system. This is done by fig. 99.

P is an ordinary Siemens relay (§ 64), whose tongue *t* vibrates between the two points *p* and *p′*. Its normal position is against the insulated stop *p*. The line current moves it against *p′*, and thus works the local circuit. The relay is similarly wound with two wires of equal length and resistance. The ends of the one wire—*the line wire*—are brought to the terminals 1 and 2, and the ends of the other wire—*the compensation wire*—to the terminals 3 and 4. If while a current is traversing the line wire a second current of exactly equal strength flow through the compensating wire in the opposite direction, the effect on the tongue of the relay must be *nothing*. The incoming line wire at A is attached to terminal 1, and the compensating wire to terminal 4. Wires from terminals 2 and 3 are connected together at *a*, and attached to the lever of the key K, which in its normal position places both these wires to earth through the back contact *k′*. The other extremity of the line wire of course makes earth through the apparatus at B, which is similarly connected up, except that the line wire is attached to 2, and 1 is connected to *a*. The other extremity of the compensating circuit makes earth at E. Resistance coils which can be varied and adjusted at will, are inserted in this circuit at R. The batteries are connected up as shown.

157. Now let the key at A be depressed; a current flows; at *a* it divides; one portion proceeds viâ 2 through the line wire of the relay, out at 1 to the line, thence to earth at B viâ 2, 1, *a* and *k′*, tending to move the tongue of the relay at A against the stop *p*. The other portion proceeds viâ 3 through the compensation wire of the relay, out at 4, to the resistance coils R, to earth at E, tending to move

the tongue of the relay against the stop *p'*. If these two currents are of equal strength they would not influence the tongue of the relay, because they tend to move it with equal force in opposite directions. But if they be of unequal strength, then the tongue of the relay will be moved in the direction of the strongest current, and by a force equivalent to that produced by a current equal to the difference of the two currents. Let us at first insert in R resistance, small compared with that of the line, then the current passing through the compensating circuit will considerably exceed in strength that passing through the line circuit. Every time the key K is depressed the relay will work, and will cause signals to be made. By gradually increasing the resistance in R the difference in strength between the two currents is diminished, until at last a point is attained where their strength is equal, and where the tongue of the relay is unaffected by the movement of the key. The artificial resistance R is now equal to that of the line circuit beyond terminal 1.

158. The line currents which are received at A from B enter at terminal 1, pass through the line wire of the relay, and out at 2, making earth through the back contact *k'*, moving the tongue against the stop *p'*, and recording signals in the usual way. Now it is evident that when A alone works to B, A's relay remains unaffected while B's relay causes the signals sent to be recorded at that station. When, under similar circumstances, B alone works to A, B's relay remains unaffected while A's relay causes the signals sent by B to be recorded at that station. But when B works to A at the same time that A is working to B, the outgoing line current from each station is reduced in strength by an amount equal to the strength of the incoming line current at each place; the compensating current at each place preponderates by the same amount. Hence marks continue to be recorded with the same force and regularity when the stations work to each other simultaneously as when they work to each other separately and independently.

159. It is evident from what has been said before (§ 154) that it is of no consequence if similar or dissimilar poles be attached to the front contact of the keys at the two stations, and hence it follows that the system can be worked as well on the double current as on the single current system. Indeed, there are conditions under which it is better to work on the double current system than on the single, and where, therefore, this mode of working is preferred.

160. There are certain irregularities in the working of such a system in actual practice which have to be provided against, due to variations in the resistance as well as in the electrostatic capacity of the line, and to what we have termed magnetic inertia. Telegraph wires, in fact, are in a constant state of change.

161. If A and B be connected together by an aërial wire supported at intervals of about 100 yards upon earthenware insulators, then the current which arrives at B from A must necessarily be less than that which leaves A because at each pole a small portion of the current escapes or leaks to earth. No earthenware support is an absolute insulator. Moisture is deposited upon its surface. The amount of this moisture continually varies, and the resistance of the insulator to the leakage of the current varies with this moisture. Hence the difference between the current leaving A and that arriving at B is constantly varying, and the effect upon the current at A is precisely the same as though the resistance of the line varied. If moisture be abundant more current leaves A, and the effect is the same as though the resistance of the line wire were reduced. If the insulators become dry, less current leaves A, and the effect is the same as though the resistance of the line were increased. In fact the resistance of the circuit does vary with the amount of moisture deposited on the insulators, and with the amount of dirt which necessarily adheres to them. Rain, fog, dew, and mist affect it. Lines exposed to the spray of the sea or the smoke of manufactories are peculiarly liable to this variation.

The resistance varies also with alterations in the physical condition of the mass of the wire, due to heat. As the temperature of a metal increases or diminishes so does its resistance. Iron wire increases in resistance 0·21 per cent. for each degree of temperature (Fahr.) through which it is raised. The diurnal variations of temperature in this climate are not great : in summer the greatest range is about 30°. This would practically not affect the comparatively short circuits used in England; but in India and America, where the circuits are much longer, and the daily variation is much greater, considerable difference is experienced in the resistance of the wire between mid-day and midnight.

162. Thus it happens that the resistance of an aërial line is constantly varying. The longer the line the greater must be the extent of this variation. The amount of variation depends essentially upon the nature and age of the line and the character of the country through which the line passes. The resistance of some lines varies in bad weather as much as 50 per cent. in one day, but remains constant in fine weather. Short lines, as a rule, are little disturbed by variations of short duration, but long lines of 200 miles and upwards are subject to constant variations due to atmospheric changes at different points. A thunderstorm here, a shower there, excessive radiation at one point, condensation at another—all tell their tale.

163. But there are other causes besides wet weather which interfere with the constancy of the resistance of a line. Earth currents and thunderstorms are a prolific source of variation. The wires are also constantly subject to accidents of various kinds, most of which produce variable resistance.

But subterranean and submarine wires are free from most of these vicissitudes. The resistance of their insulating coating is practically constant. Cables lie beneath the sea upon a cushion of equable temperature, and they are far removed from atmospheric disturbances, though they are still liable to the effects of earth's currents, and probably to those of atmospheric electricity.

164. Now what effect has this variation of the resistance of the line wire upon duplex working, and how is it provided for? Clearly it disturbs the equality of the line and compensating currents, and causes the one to preponderate over the other; and if no means were adopted to compensate for this variation, duplex telegraphy would be impossible. Therefore the compensation circuit is not only made similar to the line circuit in resistance at the outset, but it is supplied with an apparatus called *the rheostat*, which enables this resistance to be varied with the variation of the line circuit.

165. The rheostat, fig. 100, is a box of resistance coils so arranged that the motion of the handles adjusts the resistance of the compensation circuit at will. The figures over which these move indicate the number of units of resistance which will be inserted in the circuit according to the position of the handles. The arrangement is such that each handle can move over only one half of the dial: thus the maximum resistance which could be inserted in the circuit by the rheostat shown in fig. 100 would be 4400^{ω}. But as an adjunct to this a coil offering a resistance of 4000^{ω} is placed at the side of the main box, and this can be cut out of the circuit or inserted in it at will by means of the plug *p* shown in fig. 101.[1]

166. The experienced clerk knows how to vary the resistance of the compensation circuit principally by his ear. He commences by sending to the distant station, varying the resistance in the rheostat until no marks are made upon his own instrument; he then holds his key down, requesting the distant station to work to him, and varies the resistance of the rheostat until the signals received from the distant station come out not only distinct and clear to the eye but

[1] For full particulars as to the subject of Resistance and Resistance Coils the student is referred to Chap. XVI. ('The Measurement of Resistance') in Fleeming Jenkin's *Electricity and Magnetism*, and to Part II. in Culley's Handbook.

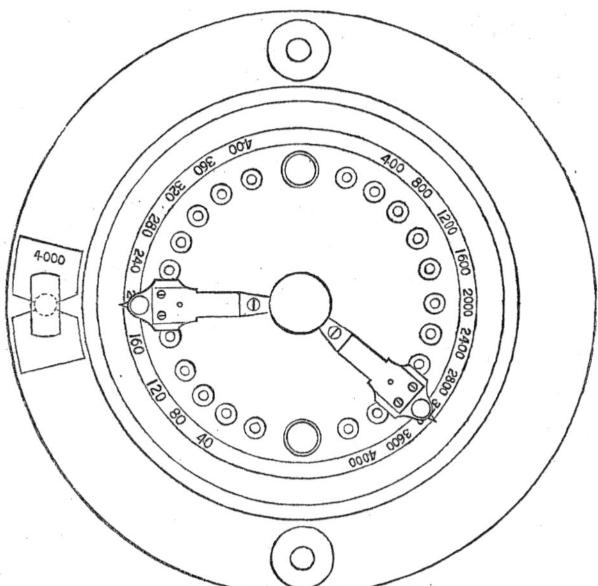

FIG. 100. Top view.

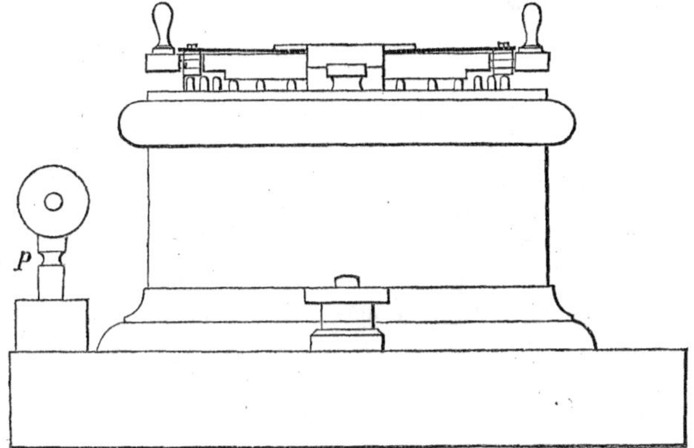

FIG. 101. Side view. ½ real size.

to the ear also. If, during working, they lose this clearness and distinctness, and appear to become *lighter*, then the resistance of the rheostat must be increased ; if, on the other hand, they become *heavier*, the resistance must be decreased. The ear is thus a true indicator of the preponderance of the one current over the other. The eye also is a sure guide, for the firmness and clearness of the signals received are soon lost if the equality of the two currents is disturbed. If the marks fail or are clipped the compensation currents preponderate, and the resistance of the rheostat must be increased ; if marks run together the line currents preponderate, and the resistance of the rheostat must be decreased.

167. The effects of electrostatic capacity have already been alluded to (§ 128), and if they were not compensated for they would seriously impair duplex working. Not only do line circuits possess this capacity, but it varies with the weather. The return currents due to induction disturb the balance. If when A is working to B (fig. 97) alone, there were induction in the line, the return currents would flow back through E and record marks. This is remedied simply by inserting in the compensation circuits condensers whose capacities are such as to produce equal effects to those of the return currents from the line (fig. 103).[1] Thus the return current from the line passes through the one wire of the electromagnet E with the same strength, and at the same time, but in the reverse direction to that of the return current from the condenser, passing through the other coil of the wire. Thus the effects of induction are neutralized.

168. But the electrostatic capacity of the line varies. It is greater in dry weather than in wet. The condensers used are adjustable to compensate for this variation ; and being inserted at certain points in the compensation circuit, their capacity can be varied in exactly the same way as the resistance of the compensation circuit.

[1] This was done by Mr. J B. Stearns, to whom alone is entirely due the successful introduction of duplex working.

169. The ear and the eye are each a guide to the want of adjustment in the condenser. If the adjustment be not right a dot will be formed at the sending station when the key is depressed if the capacity of the condenser be not large enough, and when the key is raised if it be too large. Received signals are also broken when the key is working, while they are unaffected when the key is at rest. If a continuous current be sent from the distant station an unbroken line or signal will appear when the key at the sending station is worked if the adjustment be right, but if it be wrong the line will be broken.

170. The effects of electromagnetic inertia do not introduce any irregularity in the working of the differential system. They tend only to reduce speed of working. The effects of one wire are exactly compensated by those of the other wire, so that no disturbance of signals results.

2. *The Wheatstone Bridge.*

171. We have entered so fully into the working of the differential system, that little remains to be said on the bridge

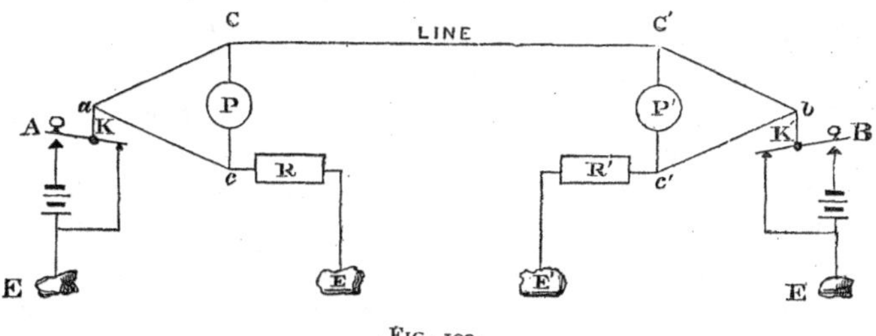

FIG. 102.

method. The differential principle is dependent on producing *an equality of currents*, but the bridge principle depends on producing an *equality of potentials*. Fig. 102 shows how the stations A and B are connected together. C C′ is the line wire, R is the rheostat, whose resistance is equal to line. *a* C and *a c* are two artificial resistances equal to each other. The key *k* is connected up with the battery in the usual way. The relay P is fixed between C and *c*. The recorder or sounder, which is worked in the ordinary way by the relay, is not shown in the figure for the sake of simplicity. The apparatus at B is precisely the same in every respect. Now when A depresses his key *k* a current is sent, and this current splits at *a*; one portion passes through the line wire to B, making earth through the apparatus at that station, and the other passes through the rheostat R to earth at A. These two currents are equal, since the resistances opposed to them are equal. Since the points C and *c* are equidistant from *a* their potentials are the same, and therefore no current can pass between them. Hence the relay is not affected when A alone is sending to B. The same exactly occurs at B when it alone is working to A. Thus we are able to send currents to a distant station without affecting our own apparatus.

172. The apparatus at B duly registers the marks when A is sending, for the line current on reaching C′ has two paths open to it—the one through C′ *c*′, the other through C′ *b*. That through C′ *c*′ works the relay, and causes marks to be recorded. The strength of this current will depend upon the relative resistances of the parts C′ *c*′, C′ *b*, *c*′ E, *c*′ *b* and *b* E. We have assumed the branches C′ *b* *c*′ *b* to be equal to each other; but it is not necessary for the resistances of the branches to be equal; they may bear any ratio to each other provided the same ratio is maintained between the resistance in R and that of the line circuit. By making the resistance of *c*′ *b* small compared with C′ *b*, we obviously increase the strength of current passing through C′ *c*′. If we made the resistance of *c*′ *b* *nil*, nearly the whole of the current would pass through C′ *c* if only the resistance of C′ *c* is small compared with that of C′ *b*; or, again, if the resistance of the branch *c*′ E is made *nil*, nearly the whole of the current would pass through C′ *c*′. But in either of these cases duplex working would be impossible, for the balance of potentials which is necessary for it depends upon the ratio of the re-

sistance in *c' b* to that in *c'* E being the same as that in C' *b* to the resistance of the line circuit.

173. To maintain duplex working we must establish a balance; that is to say, we must keep the potentials at the points C' and *c'* equal when B is working to A; hence as we vary the resistance in *c' b*, we must likewise vary that of *c'* E in the same proportion if the ratio of C' *b* to the line remains constant. But the reduction of the combined resistances in *c' b* and *c'* E has an effect upon the outgoing currents which must not be lost sight of, that is upon the current which B sends to A as well as upon that which A sends to B. The smaller the resistance is made the lower it reduces the potential of *b*, and consequenty diminishes to the same extent the strength of the current going to line. The greater portion of that sent from B's battery will take the circuit *b c'* E. A similar result would of course follow at A if the same thing were done there. Hence it is evident that the resistance of all the branches must bear a given ratio to each other in order to produce the maximum effect upon the relay at each station, and that this ratio will vary with every circuit of different resistance. It is a fact worth recording that a duplex circuit transacts *more* than the work of two wires running side by side, because such a circuit is free from the delays due to corrections and interruptions in single working, and from the disturbances and retardations due to the induction and leakage from wire to wire. Generally it may be said that the smaller the internal resistance of the battery is the more we can afford to reduce the resistance of the branches *b c'* and *c'* E, and therefore the greater will be the proportion of the current passing through the relay in C' *c'*; and the larger we can make the resistance of C' *b*, compared with that of C' *c'*, the greater will be the difference of potential between C' and *c'*, and, consequently, the stronger the current passing through C' *c'*.

The best practical results are obtained when the resistance of *c'b* is half that of C'*b*—the latter being about half that of the line. The resistance of the battery should be made as small as possible, and therefore large-sized cells are always used for duplex working.

The balance may be adjusted by altering the branch *c'b* or *c'*E or by varying both together by means of what is called *a slide*, but in practice it is found better and simpler to alter only that of *c'*E.

174. It can thus be seen that when A is working to B alone, or B is working to A alone, the apparatus at the sending station is not affected, and marks are duly recorded at B or A, as the case may be. When now A is sending to B at the same time as B is sending to A, with the resistance in the various branches at B and A duly proportioned, the equality of the potentials at the points C *c*, and at the points C' C' is disturbed; they vary, currents therefore pass, and these currents are in the same direction and of the same strength as the ordinary currents when one station alone is working. For looking first at station B when K' is depressed, the potentials at C' and *c'* are equal, provided the resistance in the various branches at B have been duly arranged; but K having also been depressed, a certain portion of the current from A reaches C'; the potential of C' is therefore increased above that of *c'*, and a current flows through C' *c'*, whose strength depends upon this difference of potential, and is manifestly the same as when K is at rest, and no current flowing from B's battery. Exactly the same reasoning will apply to A, and thus we see that while A is sending messages to B, B can also send messages simultaneously to A upon the same wire.

175. The chief point of this method which characterises it from that already described is that *it is independent of the character of the apparatus* used for receiving messages. It will work as well with the delicate mirror apparatus as with the roughest Morse recorder: by means of this arrangement the simple Sounder can be manipulated as well as the rapid Wheatstone automatic instrument. Thus no special apparatus is required for its introduction except a supply of resistance coils.

Thus duplex working is applicable to every species of instrument, and is even possible to work one instru-

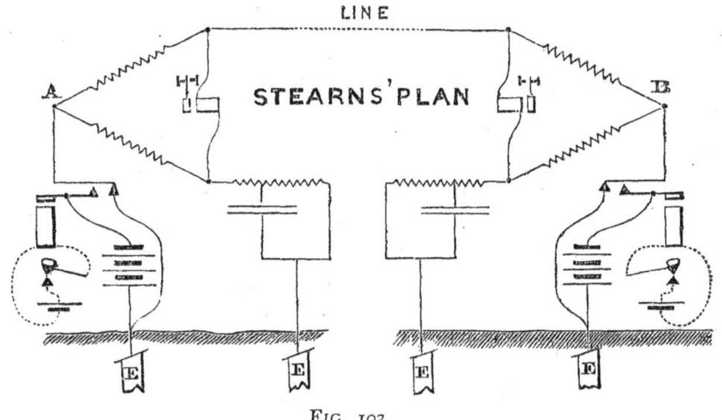

FIG. 103.

ment at one station and another kind of instrument at the other station. Thus in France Mr. Stearns has worked the Morse at one end of the line, and the Hughes at the other. Again translation is as simple as in ordinary working, and our only reason for not describing it, is the fact of its being so little used because it is so little required in England. Haverfordwest translates upon the London-Valentia duplex circuit, and has done so since July 1873, and several translating stations on one duplex circuit are employed in America.

176. The effects of a variation in the resistance of the circuit or in its electrostatic capacity are felt in the Wheatstone bridge method of duplex working as much as in the differential, and exactly the same steps are taken to obviate these in the former case as have been already described in the latter. But the irregularities due to electromagnetic inertia which are not observable in the differential system become evident in the Wheatstone bridge; and to the fact that no means have yet been adopted for overcoming these, the all but universal introduction of the differential method of working in preference to the Wheatstone bridge is probably to be attributed.

CHAPTER VI.

CONSTRUCTION.

174. TELEGRAPH lines are divided into two great classes. 1st Those in which open, that is overground, wires are employed. 2nd. Those in which covered wires, whether subterranean or submarine, are employed.

When the choice lies between these there is no hesitation whatever in selecting the former; for not only is their first cost less, but faults occurring upon them can be far more readily traced and rectified.

A. OVERGROUND LINES.

175. In England these are for the most part erected either by the sides of the roads or along the banks of the railways. Occasionally they are put up by the edge of the canals, although as a general rule the road and railway are to be preferred.

176. The advantages which road and rail respectively offer as routes for telegraph lines are so numerous that it is no easy matter to say which is to be preferred. Although the first cost of the erection of a telegraph line upon a road is greater than upon a railway, its subsequent maintenance, under certain conditions, is less. The supervision of it is likewise more perfect, for the fact that the poles are erected along the side of a road induces better inspection. The inspecting officer can hardly help inspecting every wire and insulator, and little imperfections are thus easily detected and removed before they have time to become injurious. The lineman placed in charge of a length by road must walk his length : even if he succeeds in obtaining a ride, he cannot be carried too fast for the examination of the wires. Upon the railway, on the other hand, walking is difficult, and is consequently too often neglected—the

lineman contenting himself with travelling by train, from which close inspection is next to impossible.

177. The reparation of faults again is as a general rule more speedily carried out upon roads than upon railways. In the case of the former, the lineman can start immediately the fault is reported to him : with the latter, he has not only too often to wait some time for the starting of the train, but frequently is carried past the fault, and has to return to it many miles on foot. Hence it is that not only is the number of faults less, but they are also of shorter duration, upon roads than upon railways.

It has been generally assumed that a line by road is more liable to wilful damage than one upon the railway. With the earliest telegraph lines this was true, but experience of late years does not confirm the assumption. Insulator breaking is the main evil which has been met with on roads ; but the examples made of offenders in this direction have acted as a wholesome lesson to others who may be similarly inclined.

178. *Materials.*—The materials employed in the construction of an open line of telegraph may be classed under the three following heads : 1. Supports, 2. Wires, 3. Insulators.

1. *Supports.*

179. The choice for these lies between wood and iron. In England the former is all but universally employed. The latter is occasionally introduced, but only to meet the wishes of various persons who have an idea that an iron pole is not so unsightly as a wooden one. The main advantage which wood possesses over iron for the purpose of telegraph supports lies in its first cost. In England it is about three times as cheap as iron; and although the maintenance of wood, when left in all but its natural condition, far exceeds in cost that of iron, yet timber subjected to one or other of the preservative processes which have been invented in

comparatively recent years has thus far shown so little symptoms of decay that experience does not yet warrant our forming any definite opinion as to the relative cost of maintaining it and iron. If, again, the wire by any chance touches an iron pole, good 'earth' is at once obtained by the current, and the circuit is broken down; a wire may, on the other hand, be in contact with a wooden pole, and only in very wet weather would it be found difficult to work through it.

180. The telegraph poles used in England are generally round timber, except the terminals,[1] which are square, and cut according to the requirements of the line. The round poles were at first specified to be not less than 4 in. in diameter at the top for all the minor lines, and 5 in. for all heavy trunk lines: this limit, on account of the multiplicity of wires, has now been extended to 5 in. for the former, and 6 in. for the latter. On the earliest telegraph lines the very best Baltic timber in a square form was employed, but for economical reasons this soon gave way to native-grown larch, which along with Scotch fir is now chiefly made use of. The square terminal poles are as a rule cut from Memel timber.

181. The larch and Scotch fir are felled in the months of October or November, when the sap ceases to rise, and for the sake of durability as well as strength they are required to possess the natural butts. The age of the tree when cut ranges from 25 to 50 years, and varies according to the soil upon which it is grown. Its average life as a telegraph pole—provided it has been planted in an unprepared state—may be set down at seven years, but this varies also according to the nature of its parent soil, as well as of that to which it has been transferred. If grown upon poor and chalky ground it is closer grained, and therefore more durable than if reared in rich land; whilst experience

[1] By a *terminal* pole is meant not only the last pole at each end of the line to which the wires are terminated, but also any pole at which the wires form an angle approaching to 90°.

has shown that in sandy soils it will decay more rapidly than in thick clay, in chalk than in rock, in vegetable mould than in gravel.

182. There are two kinds of decay to which a telegraph pole is liable, viz. dry-rot and wet-rot.

a. Dry-rot is very seldom met with in telegraph poles in England, and no special steps have ever therefore been taken to guard against it. It is due to a species of wood-fungus—the *Merulius lachrymans*—which destroys the tensile and cohesive power of the wood, and gradually reduces it to nothing but a fine powder. This fungus thrives best in a close moist atmosphere without draughts, such as is found in the close parts of the framing of ships. It seldom attacks open timber work except in the parts where a free circulation of air is impeded, such as the heels of iron-shod poles and the like.

b. Wet-rot is the destructive agent at work more or less on all telegraph poles, and it is to stay its ravages that all the preservative processes have been invented. This wet-rot is of two kinds, *chemical* and *mechanical.* In the first a species of slow combustion or '*eremacausis*' takes place; under the influence of heat and moisture the albuminous and nitrogenous materials of the sap ferment, and decompose the cellulin and lignin—the two constituents of which every description of timber is formed; and by a gradual process of oxidation the pole slowly but surely rots away. The germs of animal and vegetable life gradually begin to make themselves evident, and exercise a destructive influence.

But it is to the second or *mechanical* kind of wet-rot that the decay of most of the telegraph poles in England is mainly due. The point at which it makes itself evident is unfortunately that at the ground line, or, as it is more frequently termed, *the wind and water line.* It is here where the varying conditions of moisture and temperature are most felt, and to this cause the decay is undoubtedly due. For if timber be kept in a uniform state of temperature and moisture—whether it be perfectly dry or whether it be kept continually under water—a very long period will elapse before the symptoms of decay begin to make themselves evident, unless indeed dry-rot is present. But where there are rapid alternations of condition as regards heat and moisture, a process of disintegration is commenced which goes on steadily increasing until the entire structure crumbles away.

183. All the methods which have been adopted for the preservation of timber may be divided into two classes: those which have been applied *externally*, and those which have been applied *internally*.

184. The external applications are:—

a. *Seasoning.*—The poles when felled are cleared of their branches, the bark is stripped off, and the knots shaved down; they are then sheltered alike from the sun and the rain, and stacked in such a manner that the air is allowed to circulate very freely amongst them. In this way the evaporation of the sap is promoted. To get rid of the sap and the inherent germs of decay, timber has been kept immersed for a time in salt water; artificial drying in a hot-air chamber has also been resorted to, but the most perfect seasoning is that which is effected simply by exposure to free currents of air.

b. *Charring and Tarring.*—This process consists in gently roasting the butt end of the pole, after having been well seasoned, for a length of about six feet, over a slow fire, and removing it immediately the surface becomes well blackened without being burnt. The object is to expel whatever sap remains, to kill whatever animal or vegetable life may be present, to prevent absorption of moisture, when the pole is planted, by destroying the external pores of the wood, and substituting an impervious covering in their place, and finally to surround that portion most liable to decay with a powerful antiseptic in the shape of carbon. The bottom of the pole, as well as the portion which has been charred, should then be well coated with a mixture of three parts of Stockholm tar to seven parts of gas tar well boiled, and three parts of slaked lime added, care being taken to scrape off, before the tar is put on, any part of the wood which may have been burnt during the process of charring. The tar assists in more effectually accomplishing the object of the charring, and by being applied to the bottom of the pole prevents the moisture from entering there and making its way upwards under the influence of capillary action.

185. Various local applications have been made at the wind and water line to prevent decay from setting in, and although some of these have been moderately successful in attaining the end they had in view, yet all have been abandoned in favour of one or other of the preserving processes which have since been invented.

The decayed portions were at first scraped off, and asphalte was applied for some distance above and below 'the wind and water' line; cast iron and earthenware cylinders filled with asphalte were then tried without success; and finally wooden poles fitted with screw iron sockets, excluding the moisture from the wood, were put up experimentally. These last have answered the purpose on the whole well, and after the lapse of nearly twenty years the poles are reported to be in a very fair state of preservation, and to have required no renewal except in occasional instances where they had rotted near the sockets, owing probably to the latter being too deep in the ground or to the earth having been piled up over them, so as to exclude the circulation of air and admit moisture.

186. The internal applications are of two kinds: *a*. The introduction into the pores of the wood of certain metallic salts, which, by entering into chemical combination with the albuminous materials of the sap, produce chemical compounds unfavourable to decay. *b*. The introduction of some oil which, in addition to acting as an antiseptic, renders the wood waterproof.

187. Of the first class the three best known processes are: (a) *Burnetising*, (b) *Kyanising*, and (c) *Boucherising*.

(a.) *Burnetising* consists in impregnating the timber, when perfectly seasoned, with chloride of zinc. The poles are placed in an open tank filled with a solution of this salt, and allowed to remain for a length of time, varying according to the condition in which they are, until thoroughly well soaked.

(b.) *Kyanising* consists in treating the poles in exactly the same fashion with a solution of corrosive sublimate (perchloride of mercury).

(c.) *Boucherising* consists in injecting a solution of copper sulphate longitudinally through the entire length of the pole. The poles, instead of being well seasoned, must be in the green state in order to undergo this treatment; if they are at all dry the process cannot be applied. They are simply cleared of their branches, drawn into the Boucherising yard, laid upon a rack, and the solution is then applied to the butt ends under the pressure which the liquid itself has acquired from a head of about fifty feet, at which the tanks containing it are placed. This is kept on until the blue solution is observed to issue from the top of the pole. The time occupied by this operation varies according to the season of the year at which the work is carried on. It succeeds best in the spring and autumn, as at these seasons the pores of the wood are most open; frost effectually stops it. As short a time as possible should be allowed to elapse between the felling of the timber and its being placed on the Boucherising frame, for the more open the pores are the more easily is the process carried on. For this reason Scotch fir can be far more easily Boucherised than larch; quickly grown timber more easily than that of hardy growth.

188. Of the second class of internal applications the only system which requires to be mentioned is that of Creosoting. Creosote is one of the numerous products of

coal-tar, and is obtained from it by the process of distillation. When applied to timber it not only acts as a powerful antiseptic, destroying the germs of vegetable life, but by surrounding the pole with an oily covering it renders it impervious to moisture. The process can be applied only to well-seasoned timber: upon green wood it is entirely thrown away, and is in fact worse than useless, for it precludes the escape of the moisture from the pole, and thereby fosters rather than prevents the progress of decay. The poles to be creosoted are placed in a cylinder which is rendered air-tight, and from which the air is very carefully exhausted. Creosote is then applied, and forced in at a pressure varying according to the contents of the cylinder; the time during which it is applied will likewise depend upon the contents of the cylinder and the condition in which the timber is. When in proper condition timber ought to absorb eight pounds of creosote per cubic foot.

189. Of all the processes which have as yet been devised for the preservation of timber, creosoting is that which has been attended with the most beneficial results, and has given universal satisfaction. It would, in fact, be premature to assign a limit to the life of a properly-creosoted pole. Several which were erected between Fareham and Portsmouth, on the London and South-Western Railway, in 1848–49, were examined a short time since (1874) and found to all appearance to be in as good a state of preservation as on the day when they were first planted. Instances have occurred in which, after the lapse of only a few years, it has been found necessary to renew creosoted poles, but in such it is more than likely that either decay had commenced before they underwent the process, or being improperly seasoned they were not in a fit state to be subjected to it. It should, however, be added, that the creosote which at the present day is sold for the preservation of timber is inferior in quality to that which was formerly employed, and that on this account no reliable conclusions can thus far be drawn

from the experience of the past as to the value of creosoted timber in the future. The antiseptic properties of creosote are mainly due to the presence of carbolic and cresylic acids ; and these, which in former days were to be found in very large proportions in commercial creosote, have of late years become so extremely valuable as articles of commerce that they no longer exist to anything like the same extent as previously in what is now sold under that name. How far this may affect the value of the process it is at present impossible to say. Creosoted timber after being planted for a few years should from time to time be served with a coating of tar; for under the influence of the sun's rays the oil is frequently drawn forth, and either evaporates or trickles down the pole, leaving almost perfectly bleached that side which has been exposed to the sun's action. The virtue of the preserving process is thus greatly impaired, and should be restored by a coat of tar immediately the pole is observed to assume a brown rusty colour.

190. Next to creosoting the Boucherising process has found most favour, and possesses some advantages over creosoting. It can be applied to the timber immediately it is felled, and the work of preparing the pole can then be completed, without much delay, in the forest where it is cut. Boucherised timber, again, can be employed where, from objections to either the appearance or the smell, creosoted cannot. Unlike the latter, too, it may be painted, although the blue marks of the copper sulphate gradually make their appearance through the paint, unless it be of a dark green colour, which for this reason is generally preferred. To overcome this objection, so far as painting is concerned, in the case of creosote, plain poles have been served with creosote to a distance of from six to eight feet from the butt end by being submerged for a length of time in a tank of the boiling oil, and then painted for the remainder of their length. So far as experience goes this has been found to

answer very well. A great drawback to the employment of Boucherised timber is the destructive effect which the copper sulphate has upon the ironwork made use of in fitting up the pole. This being generally of a very light character, is rapidly eaten away, and in the course of a few years is rendered useless. Viewed simply in the light of a preservative, Boucherising ranks far below creosoting, and compared with it may be pronounced as a failure ; at least, so far as their application to telegraph poles is concerned. The average cost of each is about the same, and runs to nearly twopence per lineal foot.

191. Burnetising and Kyanising are seldom adopted. The latter has been abandoned mainly on account of the poisonous nature of the salt ; and the former possesses the same inherent faults as Boucherising, in addition to the further disadvantage that before it can be applied the timber must be well seasoned. Both the chloride of zinc and the copper sulphate are said to wash out, a result which is in all probability due to the compounds formed by them in the pole having lost their stability of character, owing to the entrance of either air or moisture through the pores of the wood.

192. *Iron Poles.*—As has been already remarked, iron is but little employed in England for telegraph purposes ; and when it becomes necessary to use it, the poles are usually of a light and ornamental description, to suit the wishes of those who insist upon them. But in countries where wood is extremely perishable either from natural decay or from the attacks of the white ant, and preservative processes on account of their expense cannot be introduced, as well as where the means of transport are limited, iron poles are very extensively used. On account of their weighing less than wooden poles, and being carried in pieces of convenient weight and bulk, they can be more easily conveyed from place to place.

193. The pattern of iron pole which has found the most

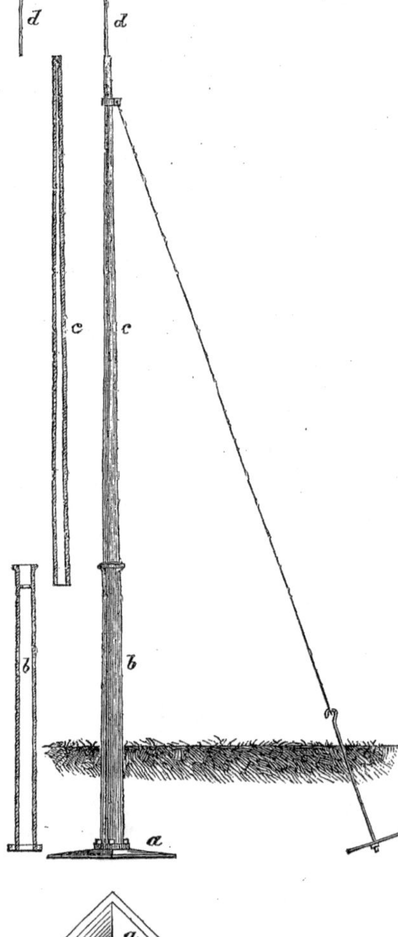

FIG. 104. $\frac{1}{45}$ real size.

general acceptance, and is now employed in almost every quarter of the globe, is the tubular post invented by Messrs. Siemens Brothers, and a drawing of which is given in fig. 104. It consists of four parts, viz.: the foot plate (*a*), the lower tube of cast iron (*b*), the upper tube of wrought iron (*c*), and the lightning conductor (*d*). The foot plate is a buckled plate of sheet-iron, of the form shown; it combines great rigidity with a certain toughness, enabling it to yield to sudden and excessive strains. The square elevation in its centre has four bolt-holes corresponding to the same number in the lugs at the bottom of the lower tube. This lower tube or socket is made of cast-iron, and is fastened to the foot-plate by means of the four bolts:

near the top it has on its inner surface a projecting iron rim, upon which the bottom of the upper portion of the pole rests. This upper portion, which is secured to the lower tube by means of a cement composed of sulphur and oxide of iron, consists of a welded wrought-iron tube tapering towards the top, to which is welded an iron ring, for the reception of the lightning-conductor.

The poles vary in size according to the work which is required of them, and as a general rule their cost may be set down at about three times that of wooden posts of the same strength. They are numbered according to their nominal breaking strain in *cwts.*

194. Another pattern of iron pole which has been largely introduced into India, where its employment for several years has given great satisfaction, and where it has been considerably modified and improved by Major Mallock, is that known as Hamilton's Standard (fig. 105).

Each standard taken separately consists of—

2 or more galvanized wrought-iron tubes *a* and *b*,
1 cast-iron socket *c*,
1 cast-iron disc plate *d*,
1 galvanized cast-iron cap *e*,
1 galvanized cast-iron lightning discharger *f*.

The wrought-iron tubes are, as a rule, each eight feet long, and are made in a series to one taper, so that two, three, four, or five may be joined together to form one post in combination with corresponding sizes of caps, sockets, &c. Each tube fits into the size smaller and on to the size larger than itself, and is strengthened by broad rings *g g*. They are put together by simple hammering.

Wooden tops were tried upon similar poles in India, but proved a failure; the timber speedily decayed before the ravages of the white ants. The insects made their way

g
FITTING LINE
f
e
b
FITTING LINE
a
c
g
g
g
g
d
d
FITTING LINE
FITTING LINE

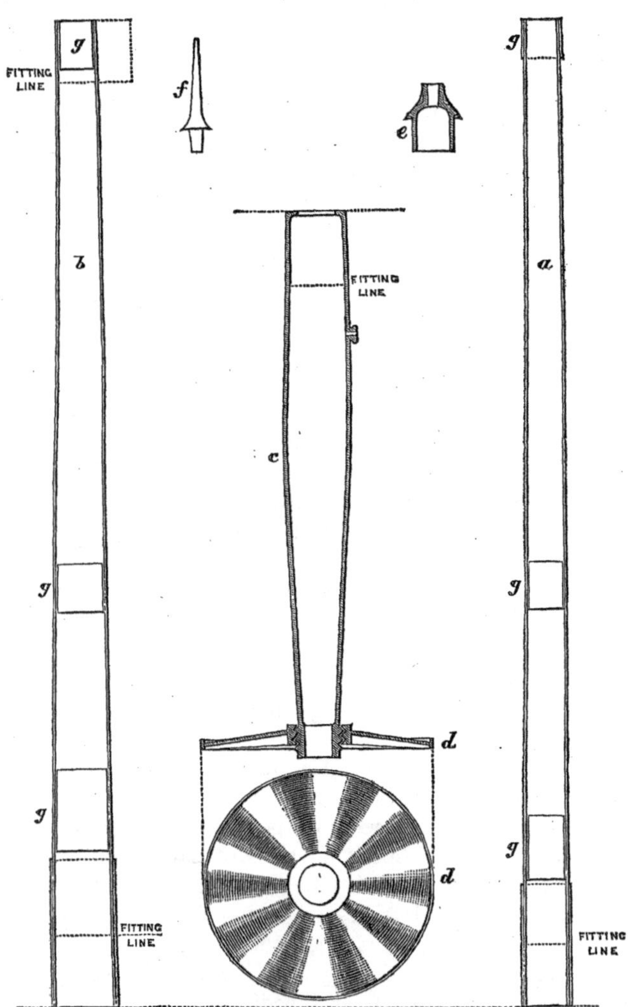

FIG. 105.

up the hollow iron tube until they reached the top, which gradually crumbled away under their attack.

195. The especial feature of the iron pole introduced by Mr. Oppenheimer, of Manchester, and extensively used in Australia, is the cast-iron base. This is shown in fig. 106. It is of the form of a pyramid, with moderately sharp cutting edges, and is driven forcibly into the ground by means of a descending weight which is worked by the tripod arrangement, shown in fig. 107. *p*, *p'* are two simple pulleys over which the ropes *r r'*, carrying the handles *h h'*, pass. These are attached to the driving weight, w, by means of the hooks, as shown. The cast-iron base is placed on the spot marked out for the pole, and has a temporary socket inserted in it. The pad or rope-buffer, *b*, is next placed over this, and over that the driving weight; a slide rod is then fitted through the head of the tripod and dropped through a hole in the top of the weight into the temporary socket. The weight is then raised along

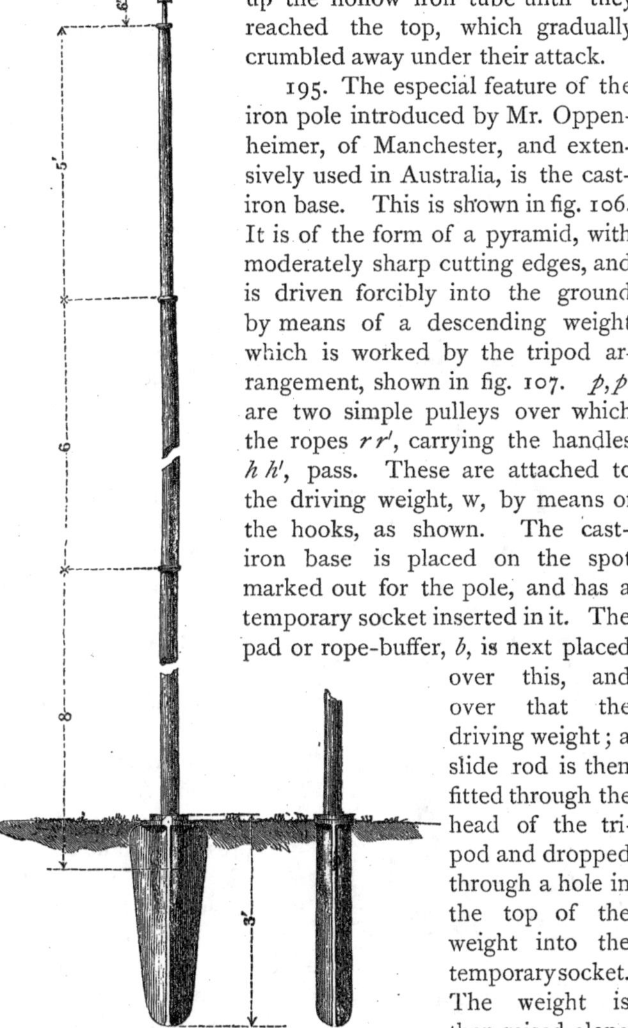

FIG 106. $\frac{1}{48}$ real size.

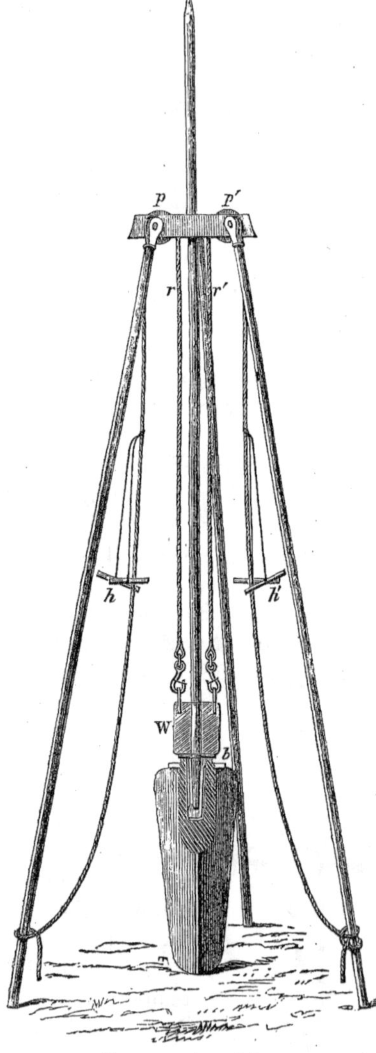

FIG. 107. $\frac{1}{45}$ real size.

this slide rod and allowed to fall on the rope-buffer; this operation is continued until the cast-iron base is level with the ground. The slide rod being removed, the iron pole is fixed in its place. This was at first effected by means of a cement, as in Siemens' pole; but the cement has been abandoned, and in its place iron wedges, varying in size, are now employed, in order to render the pole tight and secure in the socket. These poles, when erected, are very firm and unyielding at the ground line, for the earth is compressed and punned more firmly than at the outset by the base as it is driven in.

196. A pattern of iron pole patented by Messrs. Lee and Rogers of Manchester, and known as the Riband Iron Post, is shown in fig. 93. It consists of a number of straight angle irons, varying in number

according to the height and diameter of the pole. These are surrounded by a series of wrought-iron ribands (in reverse directions) riveted at each crossing in themselves, and with the angle irons. The whole is then firmly and securely attached to a circular base plate with a tripod.

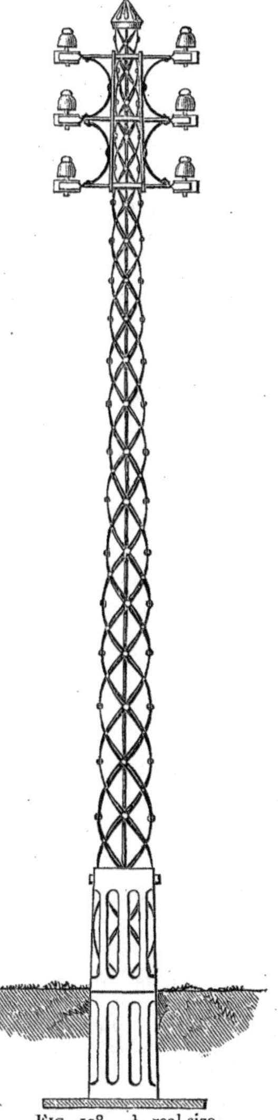

FIG. 108. $\frac{1}{45}$ real size.

This pole has not been employed to any extent for telegraphic purposes; its mechanical form is weak for lateral strains; the only advantage which it can really be said to possess over the older and better known forms of iron poles described above is in its appearance, which is undoubtedly more pleasing than theirs; but the occasions on which this would weigh in its favour are so extremely rare, that there is but little likelihood of its ever coming into general use.

2. *Wire.*

197. In selecting wire for telegraphic purposes the points to be borne in mind are economy and durability, combined with low resistance to the passage of electricity. Iron is the material which most closely fulfils these conditions, and iron wire is, consequently, all but universally

employed in open telegraph lines. Copper cannot be used, not only on account of its intrinsic value, but from the fact that it bears little tension, stretches excessively, soon loses its elasticity, and is much affected by temperature: it has been tried in different countries, but has always proved a failure.

198. *Iron Wire.* There are various qualities of iron wire, known under the names of 'best,' 'best best,' 'extra best best,' and 'charcoal' wire.[1] The last named is the most expensive, and in the earlier telegraph lines it was in general use on account of its being more easily welded than wire of a lower quality. To the present day it is largely employed in France and elsewhere. The improvements which during the last twelve or fifteen years have been made in the manufacture of iron wire—getting rid of welds altogether—led to the abandonment of charcoal wire and the substitution in its place of the quality of soft iron wire known as 'best best,' which is now all but universally employed in England for telegraph work. For exceptionally longer spans a harder iron wire, or occasionally steel wire, which is stronger, is used.

199. The 'pigs' of iron from which the wire is drawn are first of all 'puddled' in a furnace; the ball of puddled iron is then placed under a very heavy hammer, by which it is beaten out into a compact form. It is then passed between a series of rollers, from which it finally emerges in the shape of a bar, much increased in length and reduced in thickness.

200. The bar is then passed through what is technically known as 'the rolling-mill.' This machine consists of a series of rollers, placed in pairs alternately horizontal and

[1] These terms are employed simply to distinguish the various qualities of wire. 'Best' is the ordinary puddled wire, and is in fact indiscriminately applied to almost any kind of telegraph wire or iron bar. 'Best best' indicates a quality superior to the former, and in its manufacture more expensive pig-iron is used. 'Extra best best' is a still higher quality, and is obtained by the introduction of charcoal iron in connection with the last named.

vertical. Each is grooved, but the size of the groove diminishes with each succeeding pair of rollers. Thus as the bar passes through these its length is increased and its diameter reduced to whatever extent may be desired. Each pair of rollers have their speed controlled by separate driving gear, and upon reflection the necessity of this will be seen: the length of the bar increases between every pair, so that what enters the mill at a very slow pace finally issues from it at a tremendous rate of speed, varying of course according to the diameter to which the bar has to be reduced to convert it into wire.

A wire is reduced to a smaller gauge by being forcibly drawn through a 'die' whose diameter is rather less than that of the wire itself.

201. The largest wire employed in England is that known in the Birmingham wire gauge as No. 4, having a diameter of ·240 inch. It is used, however, only under exceptional circumstances, or upon very long circuits. The wire in general use for all through circuits is No. 8 (diameter ·170 inch). No. 11 (diameter ·125 inch) is used for all short circuits of minor importance. For binding purposes No. 16 (diameter ·065 inch) galvanized wire of the best selected charcoal iron, highly annealed, very tough, soft and pliable, is employed. It is the only form of charcoal wire still retained in England.

202. Iron wire if left unprotected speedily becomes oxidised in the open air, and for this reason ought to be covered with a protective covering of zinc, which is not acted upon by either moist or dry air. To this process of coating the wire with zinc the term *galvanizing* is applied, and wire so treated is known as galvanized iron wire. Of late years considerable improvement has been made in carrying this out; the most approved method is the following, which combines into one the three processes of annealing, cleaning, and galvanizing the wire:—The hard iron wire is first tempered by being passed through a heated tube; it is

then drawn for a few seconds through a bath of hydrochloric acid, which serves to remove all the surface impurities; it is next guided by means of rollers through a bath of molten zinc. After leaving this the wire passes through a mass of different material—including sand, &c.—which acts as a gentle scraper, and is finally wound on the coiling drums in a thoroughly galvanized state.

203. The wire should be manufactured in as long lengths as possible, consistent with convenience in handling it when being erected; for, in the great majority of cases where wires break, from any other cause than that of being damaged during their erection or chafed after they are up, it will be found that the breakage occurs at a *weld.* There should be no weld in properly manufactured wire.

204. Flaws in a newly erected wire, due to impurities in the shape of cinders, &c., which have been allowed to find their way into it during the process of manufacture, will make themselves evident on the occasion of the first frost, by the wire breaking at the points where they exist. For this reason the bars from which the wire is drawn should be carefully selected from the best material only, and the danger may be still further obviated by using, instead of one solid bar, a mass of metal composed of several different pieces laid together. Thus if eight pieces of iron be piled together—say four $1\frac{1}{2}$-inch billets boxed up in 5-inch tops and bottoms with 3-inch sides—and if these be well washed, heated, and rolled out into bars of about $1\frac{1}{2}$ inch diameter, they will when passed through the rolling-mill produce an entire length of about one-third of a mile of the No. 8 gauge. Iron wire manufactured in this way is found to combine the ductility of strand with the homogeneousness of solid iron, and reduces to a minimum any danger of breakages occurring through flaws in it.

205. It is essential that the wire employed in telegraphy should be free from flaws, welds, and impurities—and that its tension or power to resist strain should

be uniform throughout. Both these objects are attained by one and the same process, which is as follows:—The galvanized iron wire is placed on a simple loose wheel, or 'swift,' as it is technically termed; from this it is drawn alternately over and under three or more small pulleys arranged in the manner shown in fig. 109. It then passes

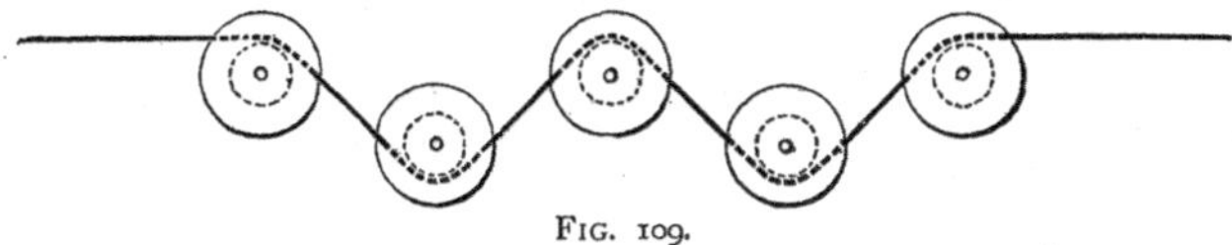

Fig. 109.

round a large V sheave, and is finally wound upon a drum which is turned with a velocity greater by about two per cent. than that of the V sheave. The strain, which it will be seen is thus put upon the wire, not only tests it, but makes evident any defects which it may contain. To detect these the coils of wire previously to being issued should be carefully examined by the eye.

206. There are four mechanical tests to which the galvanized iron wire may be subjected in order to prove its quality. 1st. It should be capable of being bent backwards and forwards at right angles to itself a certain number of times without breaking. 2nd. It should be capable of being wound around itself a certain number of times without showing signs of splitting. 3rd. It should be able to bear a certain number of twists, varying according to the length, without splitting; this is the ductility test usually employed. 4th. It ought to be able to carry a certain weight or resist a certain strain without breaking.

The last named is the test which is generally employed. It is carried out either by means of a hydraulic machine or by a scale and weights. The objection to the former method is that the additional strain in testing is too rapidly applied, and the wire not having had time to yield to the previous strain will really show a higher power than it actually possesses; with the latter the additional weights should not be put on until the wire has been allowed ample time to stretch under the influence of those previously in the scale.

207. Good soft iron wire, such as that employed in telegraphy, should be able to bear from 18 to 20 per cent. elongation. The breaking strain will of course vary according to the gauge of the wire; that for No. 8 should not be lower than 1,200 pounds, and for No. 11 not less than 650 pounds. Galvanizing, although it does not seem to have any appreciable effect upon the breaking strain of the wire, hardens to some extent the iron, and thus diminishes the coefficient of elongation.

208. As the result of a very large number of experiments we subjoin a sliding-scale, which will serve to demonstrate what tests the various qualities of iron wire should bear. It will be observed from this that any falling off in ductility—which is indicated by the decrease in the number of twists or turns that the wire is capable of bearing – must be compensated for by increased breaking strain :—

Sliding-Scale for Twists in Six inches, and minimum Breaking Strain.

Gauge	Diam. inches	Breaking Strain	Twists	Breaking Strain	Twists	Breaking Strain	Twists	Breaking Strain	Twists	Breaking Strain	Twists	Average Elongation
I. *Best Best.*												
No. 4	·240	2300	11	2350	10	2400	9	2450	8	2500	7	15 per cent.
5	·220	2000	12	2050	11	2100	10	2150	9	2200	8	
6	·200	1700	13	1750	12	1800	11	1850	10	1900	9	
7	·180	1340	14	1380	13	1420	12	1460	11	1500	10	
*8	·165	1080	15	1110	14	1140	13	1170	12	1200	11	
9	·150	830	16	860	15	890	14	920	13	950	12	
10	·135	725	17	750	16	775	15	800	14	825	13	
†11	·120	575	18	600	17	625	16	650	15	675	14	

* This is ·05 inches smaller in diameter than the No. 8 wire used by the Post-Office Telegraph Department in England.

† This is ·005 inches smaller in diameter than the No. 11 employed by the same Department.

Sliding-Scale for Twists in Six inches, and minimum Breaking Strain—continued.

Gauge	Diam. inches	Breaking Strain	Twists	Breaking Strain	Twists	Breaking Strain	Twists	Breaking Strain	Twists	Breaking Strain	Twists	Average Elongation
						II. *Extra Best Best.*						
No. 4	·240	2300	13	2350	12	2400	11	2450	10	2500	9	16 to 18 per cent.
5	·220	2000	14	2050	13	2100	12	2150	11	2200	10	
6	·200	1700	15	1750	14	1800	13	1850	12	1900	11	
7	·180	1340	16	1380	15	1420	14	1460	13	1500	12	
* 8	·165	1080	17	1110	16	1140	15	1170	14	1200	13	
9	·150	830	18	860	17	890	16	920	15	950	14	
10	·135	725	19	750	18	775	17	800	16	825	15	
† 11	·120	575	20	600	19	625	18	650	17	675	16	
						III. *Charcoal.*						
No. 4	·240	2300	15	2350	14	2400	13	2450	12	2500	11	18 per cent.
5	·220	2100	16	2150	15	2200	14	2250	13	2300	12	
6	·200	1700	17	1750	16	1800	15	1850	14	1190	13	
7	·180	1300	18	1350	17	1400	16	1450	15	1500	14	
* 8	·165	1080	19	1110	18	1140	17	1170	16	1200	15	
9	·150	830	20	860	19	890	18	920	17	950	16	
10	·135	725	21	750	20	775	19	800	18	825	17	
† 11	·120	575	22	600	21	625	20	650	19	675	18	

209. These are the main points worthy of mention in connection with the subject of galvanized iron telegraph wire. The following are the conditions applicable to the supply of No. 8 manufactured for the Post Office Telegraph Department in England by Messrs. Johnson and Nephew of Manchester[1]:—

1. The wire must be of the gauge known as No. 8 Birmingham wire gauge (diameter ·170 of an inch).

2. The wire to be highly annealed and very soft and

* This is ·05 inches smaller in diameter than the No. 8 wire used by the Post-Office Telegraph Department in England.

† This is ·005 inches smaller in diameter than the No. 11 employed by the same Department.

[1] The various processes in connection with the manufacture of iron wire, described above, are those which are carried on by this well-known firm.

pliable, and to be galvanized. The wire is not required to possess great tensile strength, but must be capable of elongating 18 per cent. without breaking after being galvanized.

3. The wire to be entirely free from scales, inequalities, flaws, splits, and other defects, and to be cylindrical.

4. No deviation greater than ·005 of an inch either way from the prescribed diameter will be allowed.

5. The whole of the wire to be passed under and over three or more pulleys or fixed studs placed in a position similar to that shown in fig. 109, and such as shall, in the opinion of the engineer of the Postmaster-General, admit of the quality of the wire as regards freedom from splits being sufficiently tested.

6. The whole of the wire to be stretched 2 per cent. by machinery, and after being stretched to be coiled carefully so as to contain no bends or indentations, but in all respects to resemble newly-drawn wire.

7. If during the process of testing the wire between the pulleys or fixed studs, or if during the process of stretching it, more than 5 per cent. of the bundles break, crack, or show any defect, the whole of the broken bundles to be rejected. If not more than 5 per cent. prove defective the whole of the broken bundles will be accepted, provided always that no piece in any bundle be less than 40 pounds in weight. The makers are not to weld, join, or otherwise splice any wire that may break or prove defective, but deliver accepted wire as it comes from the testing.

8. The wire to be made up in coils weighing not less than 90 lbs. and not more than 120 lbs. each, and to consist of one piece or length only (see, however, preceding clause), which must be warranted not to contain any weld, joint, or splice whatsoever, either in the rod before being drawn or in the wire after it is drawn.

210. *The Compound Telegraph Wire.*—This, an American invention of comparatively recent date, has been employed

to a considerable extent in the construction of telegraph lines in various quarters of the globe. It consists of a tinned steel wire of the best quality, around which a copper strip is wound helically. The steel and copper having been drawn through a draw plate, are firmly soldered together, and the compound wire thus produced combines lightness with strength, and a far higher conductivity than iron wire of the same gauge possesses.

There can be no question that, could such a wire as this compound telegraph wire be made to last, it would be largely adopted, especially in countries where the transport of materials is expensive; and even in other countries the saving effected by its use in the other materials employed, as well as in the labour in the construction of a telegraph line, would form a very considerable item, and probably warrant its adoption in preference to iron wire.

In addition to this, the induction between wire and wire —which so often interferes with the working of fast-speed instruments on long circuits—would be less felt with a small wire offering the same electrical resistance as one much larger, and a powerful argument in its favour might be drawn from this fact. But until the indispensable condition of durability can be ensured, it is almost idle to discuss the merits of the Compound Telegraph Wire.

211. *The Wire Gauge.*—Before quitting the subject of iron-wire it is desirable to draw attention to the gauge according to which it is specified. Until quite recently this has invariably been what is known as the Birmingham Wire Gauge. This wire gauge, however, varies with every manufacturer, and there is not only no standard in existence from which it can be corrected, but, from the fact that the basis on which it was originally formed is hid in obscurity, it seems impossible to have one reproduced in a reliable shape. Various suggestions have from time to time been made for the introduction of a new wire gauge which should be coherent throughout, be based on a regular increasing

series, and start from some recognised and well-known unit. It seems at first sight difficult to decide whether a gauge of this description should be based upon *weight* or diameter. But when it is remembered that wire is purchased, transported, and distributed along the line by weight, that its breaking strain is in proportion to its weight, that its electrical resistance—varying as the square of its diameter—is a function of its weight; and finally that weight is invariable in all temperatures and latitudes, it will be admitted that multiples of a unit of weight are the natural telegraph wire gauge. A size of wire dependent upon a number of pounds per mile will be constant as long as pounds and miles exist, and if these units are adopted as a basis there is a ready means of correcting the gauge at all times.

At the suggestion of Major Mallock, the Director of Construction, the Government Telegraph Department of India has adopted an iron wire gauge of this nature. It is based upon the weight per mile of the wire, and the unit is a wire weighing twenty-five pounds per statute mile;[1] all other sizes of wire are known by their multiples of this unit, and in terms of this unit size, the resistance, breaking strain, and comparative strain of the wire upon the insulators or posts, can all be readily calculated.

3. *Insulators.*

212. In the manufacture of the insulators two points have to be kept in view. 1st. The material. 2nd. The form.

1. *The Material.*—The main object of course in the selection of this is to find a substance which will offer the greatest possible resistance to the passage of electricity. Nothing has yet been found which will perfectly insulate; nor can a theoretically perfect body in this respect ever be looked for. Porous substances are inadmissible on account of their absorbing moisture too readily, and being thus transformed into conductors. A glaze or surface can be im-

[1] This is the No. 19 of the Birmingham wire gauge.

parted to them, but recourse should never be had to this; only upon bodies which are in every respect suitable should a glaze be put, and then for the purpose of forming a fine hard surface. A smooth hard surface is indispensable; with it there is no danger of the wire being worn through by friction, nor can dirt and dust adhere to the insulator so firmly as not to be washed off by a good shower of rain.

213. *Glass* possesses both of the qualifications named above, viz., high resistance to the passage of electricity and a smooth hard surface; but along with these it has one inherent disadvantage which is fatal to its employment as an insulator. It is a very hygroscopic body—that is to say, it condenses the moisture from the air very readily, and in a climate such as that of England it is for this reason altogether unsuitable. The surface of a glass insulator will be almost always covered more or less with a thin conducting film of moisture. It is moreover very brittle, and was consequently abandoned in favour of one or other of its rival competitors. Of late years, however, Mr. Brooks has introduced in America a form of insulator which is manufactured from blown glass, and is stated to have given very good results. These he considers to be mainly due to the 'air surface' of the insulator, nothing but dry air being allowed to come into contact with it whilst it is being manufactured.

214. *Ebonite*[1] is another substance which possesses many good points to recommend it as an insulator. It offers a very high resistance; it is strong, and when first used possesses a good smooth surface; it has an unassuming appearance, and so escapes from wilful damage, where glass, porcelain, &c., owing to their inviting look, would run the risk of being broken. The great defect which ebonite labours under, and that which practically precludes its employment as an insulator when exposed to the weather, is the fact that

[1] Ebonite is a mixture of two or three parts of sulphur and five parts of caoutchouc baked for several hours at 170° F. under a pressure of four or five atmospheres.

its surface deteriorates rapidly. Instead of remaining smooth and hard as when the insulator was first erected, it gradually becomes porous and spongy; dirt and moisture form upon it, and so deprive the insulator of one of the first qualities which it ought to possess.

215. *Porcelain* has been and is still largely employed in the manufacture of insulators. Its insulating power is high; it possesses a good smooth surface; and provided it has been perfectly vitrified throughout so as to be homogeneous, impervious to moisture, and free from flaws, it is eminently adapted for the formation of an insulator. Porcelain, however, varies very much in its quality; and unless the manufacture has been thoroughly carried out, with the greatest care, no reliance can be placed upon it. To all kinds of porcelain a glaze can be communicated; and so long as this remains good, so long will the insulator continue to give good results; but immediately the glaze cracks, which it soon does, moisture enters the mass if it is porous to any extent, and the value of the insulator is greatly diminished. Unfortunately there is no means of testing the manufacture when the insulator is new and the glaze has been imparted to it, except by breaking it, and the integrity of the manufacturer has therefore to be solely relied upon.

216. *Brown Earthenware* is the material from which most of the insulators employed at the present day in England are formed. It does not insulate so highly as good porcelain, nor can it be so perfectly glazed, but it is produced with greater uniformity of quality, and its manufacture can be more thoroughly relied upon. It possesses the further advantage of cheapness over the materials which have been already named.

217. 2. *The Form.*—Equally important as the material of which an insulator should be composed is the form which should be given to it. In considering this the main object to be kept in view is the same as in the selection of the material, viz. the highest possible resistance to a leakage of the cur-

rent; at the same time the strength of the insulator as a support must not be altogether lost sight of. Seeing, however, that the insulators have little more than the weight of the wire to withstand, except at the terminal posts, no trouble is experienced in suiting the form of insulator to this. The main difficulty which has to be surmounted is the leakage which takes place more or less at every support; every insulator is to a certain extent a fault, and the magnitude of the fault depends upon the form which the insulator possesses. The passage of a current of electricity through the wire depends upon the gauge of the latter; it is a function of its mass. The same cannot be said of the insulator; the resistance to the path of the current, instead of depending upon the mass of the insulator, is mainly a question of surface. The most perfect form of insulator will be that in which the surface exposed is a minimum, and the wire is as far as it can be from the insulator's support, due allowance being of course made for the insulator itself being sufficiently strong.

218. Numerous forms have from time to time been tried; that which is in general use in England with the brown earthenware is shown in fig. 110. It consists of two distinct and separate cups, *c* and *c'*, which are fitted into each other by means of a cement composed of equal parts by weight of fine pit sand, Portland cement, and plaster of Paris. Into the inner cup *c* a galvanized iron bolt *b* is inserted and fixed by means of a cement composed of

5 parts by weight of clean river sand, sifted.
3 „ „ engine ashes (from an old engine fire-box).
2 „ „ pine resin.

A groove is cut on the surface of the outside cup, and into it the line wire is placed and firmly bound, as will be afterwards explained. This form of insulator combines several advantages, to which attention may briefly be drawn. Whatever current escapes from the wire in the groove must make its way over the entire surface of both cups before

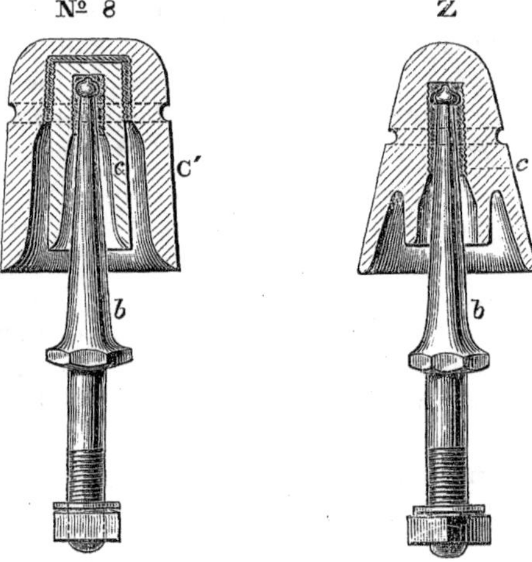

FIG. 110. ¼ real size. FIG. 111. ¼ real size.

it can reach the bolt. By having two cups again greater reliance can be placed upon the quality of the earthenware; the two small pieces can be better burnt and vitrified than one larger portion; and the probability of flaws or faults occurring in such cup is very remote. At the same time by means of this arrangement one portion of the insulator is open to the cleansing action of the rain which serves to remove any dust or dirt apt to adhere to it; whilst the other is kept dry, and in wet weather continues to offer considerable resistance to any escape of the current.

219. Another form of insulator employed in England on minor lines is that known as the 'Z' insulator, and shown in fig. 111. It closely resembles the form already described, and differs from it in having a double cup *c* all formed of one piece of earthenware instead of two separate cups fitted into each other.

220. When the insulators have to be protected from either wilful or accidental damage, such as that occasioned by stone-throwing and the like, it is customary to cover them with an iron cap, and bind the wire into a small lug upon the surface of it. The inconvenience attending the use of iron caps is occasioned by the accumulation of dust, insects, &c., beneath the hood; the iron cap protects them from the cleansing influence of the rain, and so leads in time to a deterioration of the insulation. An effort has been made to get over this by cutting slits in the iron cap, and although this has to some extent remedied the evil, yet only where actually rendered necessary by either of the causes named above should iron-capped insulators be had recourse to.

221. Great difficulty is invariably experienced in preserving the insulation upon those lines which skirt the sea-coast, no matter what material is employed or what form of insulator is adopted. The insulator becomes coated with salt, which being more or less moist, conducts in all except the driest possible weather. The difficulty is greatly increased when the wind is from the sea. Upon no account should iron-capped insulators be made use of upon such lines as these; advantage should be taken of the rain to the utmost for washing the salt from off the outside surface at least of the outer cup; it materially improves the insulation. Wire covered with tarred tape or hemp is occasionally employed in extreme cases of this nature; but by chafing against the insulator the tape or hemp gradually gets rubbed off, and leaves the wire exposed just at the point where protection of this nature is most required. Open wires skirting the sea-coast should therefore be resorted to only when no other route by which they might be carried is available.

CHAPTER VII.

CONSTRUCTION (*continued*).

222. *Surveying.*—The route for a line of telegraph, whether by road or by rail, having been decided upon, the next point is to make a careful survey of it. For this purpose the surveying officer should be provided with a book prepared upon the following plan, in which the requisite particulars are inserted to enable him to estimate the total quantity of stores which will be required, and to provide for their being laid out, as well as to make arrangements for obtaining the permission to erect the poles where such is required.

SURVEY of from to .

1	2	3	POLES		6	
Number of Pole	Span	Distance from	4 Length	5 Quality	Stay or Strut	REMARKS as to Consents, Obstacles, &c.

This schedule explains itself; yet it may be well to state that:—

Column 1 contains the consecutive number of each pole, from one terminal station to the other.

Column 2 contains the span, or distance between each successive pole.

Column 3 gives the distance from some fixed object; the fence or ditch, for instance, in the case of a road line, or the metals for a line constructed upon the railway. This is to guide the hole-diggers in the event of the marks, made by the surveying officer to indicate the position of the pole, having been removed.

Column 4 gives the length of the poles.

Column 5 explains the character of the pole required, which will depend upon the work which it has to do. O is used to represent the pole of ordinary scantling, S a stiff one, and V S stands for a very stiff pole. If A poles or double poles are necessary, this is stated in this column.

Column 6 states whether a stay or strut is required, whether single or double, and should give the strength or number of wires necessary for the stay, and the length of the strut.

Column 7 is intended to provide for any additional particulars which may be deemed necessary, such as shackling, crossing roads, noting any obstacles in the way, the names and addresses of those who will have to be consulted in the matter of way-leave, &c.

223. The surveying officer should have at least two assistants. They should carry with them a supply of wooden stakes, and a can of white paint, to mark the position of the poles, and they should likewise be provided with three conspicuous white rods. The latter are indispensable if an accurate survey is to be made, especially for a line carrying more than one wire, for only by means of them can an estimate be formed of the amount of curve, and consequent strain, to which each pole will be subjected; and without this information the requisite provision for suitable timber and proper staying or strutting cannot be made. The positions of the poles may be marked in various ways: the plan which has been found to answer best is the insertion of wooden stakes in the ground, aided by a distinctive mark of white paint, to provide against their being removed.

224. *Hole-digging.*—The operation of digging a hole for a telegraph pole, although to all appearance simple enough, is yet more difficult than at first sight would be imagined. The holes should invariably be dug in the line

of the wires, as at A, and never at right angles to it, as at B (fig. 112), the object being to get the solid natural earth as

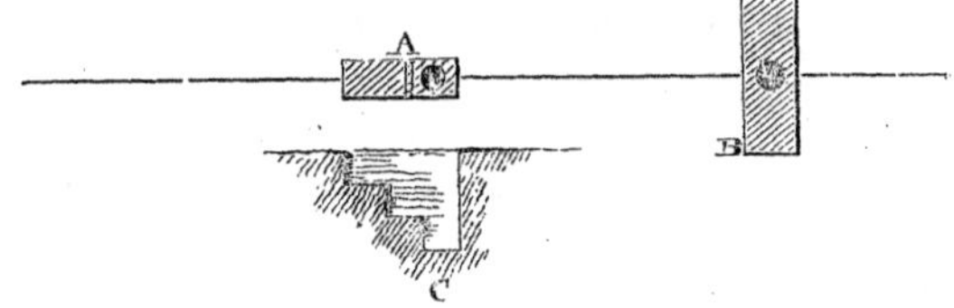

FIG. 112.

much as possible against the lateral strain of the wires. The rectangular opening which is thus made averages about four feet in length by two in width : this size is continued to a depth of about two-and-a-half feet below the surface, whence, by a step-like arrangement, the length of the opening is gradually curtailed, until at the bottom it does not exceed one foot, as shown in C.

225. As little of the ground as possible should be disturbed, for no matter how well the punning and ramming may be done after the pole is planted, yet a considerable time will always elapse before the earth settles back to its former condition, during which the pole is less able to support any strain that may be put upon it.

For this reason various tools have been devised, whose object is to remove only just sufficient earth to admit of the pole being planted; and which, in addition to effecting this, combine several other incidental advantages of considerable value. When it is borne in mind that in order to dig a hole four feet six inches deep for an ordinary telegraph pole, by the pick and shovel in the usual method, no less than thirty cubic feet of soil, representing a weight varying from 2850 to 3,600 pounds, according to the nature of the ground, have to be removed, whereas not more than three and a half cubic feet, or about 376 pounds, need actually be disturbed, it will readily be seen how much room there is for improvement in this branch of telegraph construction.

226. One of the earliest efforts made in this direction was in Spain, where a tool, since known as the *Spanish Spoon*, was devised. Various modifications have from time to time been introduced, but they are all constructed on the same principle, which is that shown in fig. 113.

It consists of a segment of a metallic disc *a*, the chord of which serves as a cutting edge. The periphery is fitted with a ledge *c* two inches in height, which serves to retain the accumulation of the soil upon it. The whole is fitted to a wooden handle *b*. The adjunct to the spoon is a long bar, by means of which the soil is first loosened: the spoon is then inserted, and to it a rotating motion is conveyed, by which the earth is heaped up: the whole is then removed, and the bar again employed. For light lines, on which the poles need not be inserted to a greater depth than four feet, the *Spanish Spoon* answers the purpose for which it is intended very fairly; but in heavy lines, where holes varying from six to seven feet in depth are required, it cannot be pronounced a success. The difficulty of loosening and collecting the soil increases to a very great extent with the depth, and the advantage which at the outset it possesses over the pick and shovel in point of speed is almost, if not entirely, lost before a six or seven foot hole is completed.

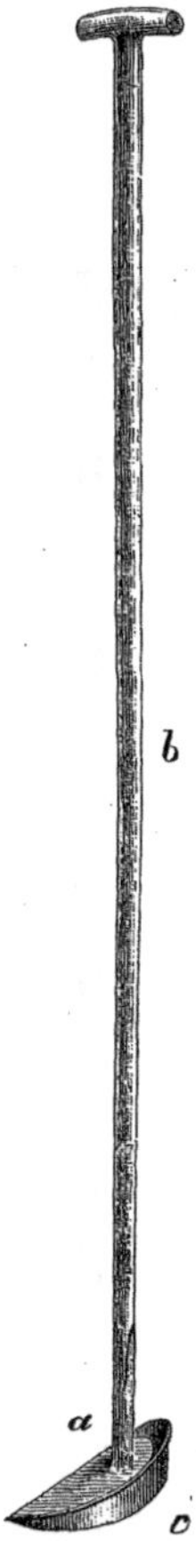

FIG. 113.
1/18 real size.

227. The introduction of *Earth Borers* of late years, for the purpose of excavating the holes for telegraph poles, is the most important point in connection with this branch of the subject. Various kinds have been tried, but those which are most generally known are the inventions of Spiller, Bohlken, and Marshall.

Spiller's is but a modification on a large scale of the ordinary ship's auger, which is forced into the ground, and in clay or sand has been found to work well.

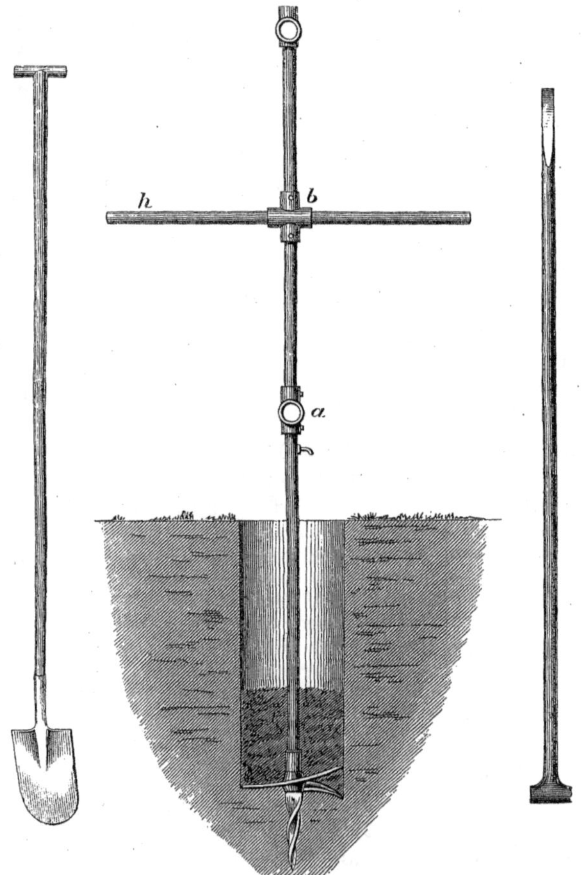

FIG. 114. $\frac{1}{24}$ real size.

228. Marshall's borer, although resembling Bohlken's,

has several distinctive features about it. A section of the apparatus is shown in fig. 114. The cutting blade, a plan and section of which is given in fig. 115, consists

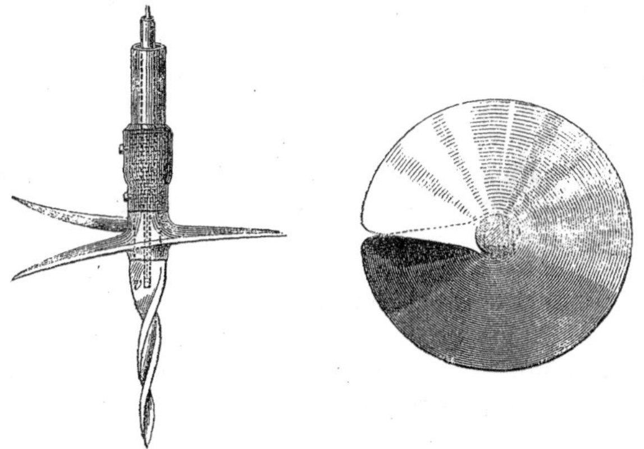

FIG. 115. $\frac{1}{15}$ real size.

of a metal disc cut from the centre to the circumference, and having the two edges bent into the V shape shown. The lower forms the cutting edge, and as the apparatus is rotated, the earth passes through the radial V opening on to the upper surface of the blade, from which it is removed by extracting the apparatus from time to time from the ground. The stock to which the blade is attached is squared at the end, and has screwed on to it a tapering metal point, which, in addition to serving as a nut, plays the part of a drill in front of the cutting plate, and so facilitates its work to some extent. The stock is attached to two or more sections of tube, according to the depth of the hole: these are provided with cross sockets as shown at *a* and *b*, in fig. 114, to admit of the insertion of the handle *h*, which is employed for rotating and lifting the apparatus.

229. The 'punner bar' should invariably accompany Marshall's borer, and forms in fact an essential feature in his

arrangement. It is shown to the right in fig. 114. The upper end of this is tapered down to the form of a chisel, with the point tempered to deal with stones, and is used for loosening the soil as well as breaking and removing as far as possible whatever obstacles are met with in the hole: the lower end, which is shaped like a punner, is employed for ramming and consolidating the soil around the pole when once planted. The borer is rotated by two men walking steadily round and pushing the handle before them : at intervals it is lifted, and the earth removed : the chisel end of the punner bar, if need be, is inserted, and the process repeated. A shovel attached to a long handle, and shown to the left in fig. 114, should likewise accompany the apparatus. In sandy or gravelly soils it is employed to remove the loose earth which does not adhere to the blade of the borer. An evident drawback to the general employment, even in suitable soils, of a borer of this form, is the impossibility to work it by the sides of fences, where, in road telegraphy, poles have generally to be placed.

Another drawback is the enormous strain thrown upon the men when lifting a load out of the hole, especially when some considerable depth has been obtained : for in a clay soil, or if the ground is close, not only is there the weight of what has accumulated on the plate to be lifted, but that of a superincumbent column of air as well. This difficulty was got over by inserting a small valve *v* (fig. 115), which can be opened at will by the workmen, and greatly facilitates the raising : the best cure, however, is perhaps to raise the borer more frequently, and not to wait for such heavy loads upon it. The difficulty in raising heavy poles so as to let them slip into the holes which have been prepared for them is a decided disadvantage inherent to the employment of all earth borers : light poles can be handled easily enough, but the same cannot be said when poles from thirty to forty feet in length or even more, have to be dealt with. The only possible way of lifting them is by means of shears, which have to be carried about with the gang of workmen employed; and although the work can then be performed with comparative ease, the multiplication of tools is always a disadvantage more or less, and, in countries where roads do not exist along the routes of the telegraph lines, should be avoided to the utmost.

230. This borer has, on the whole, given very satisfactory results in England, notwithstanding the *vis inertiæ* of the workmen, which, in common with almost every new invention, it has had to encounter: in rocky soil it is of course all but inadmissible, and can do no more than the pick and shovel there; but in most instances it can be employed wherever they can be, and generally at a considerable saving of expense.

In addition to this diminution of labour and expense, and the greater stability which a line erected by the borer unquestionably possesses, the apparatus can claim the further advantage of being workable without baling or pumping in wet soils, which are such a fruitful source of trouble to the workmen with the pick and shovel; and the comparatively clean job which it makes owing to the small amount of material which is disturbed may be advanced as another argument in its favour; especially in such a country as England, where private rights have so frequently to be encroached upon.

231. In the latest form of borer, brought out by Marshall, the tapering metal point is entirely dispensed with, and the cutting-plate itself is in the form of a screw, and thus acts both as a drill and cutting-plate. This apparatus is cheaper than the earlier issue, and for light work can be worked by one man. Beyond this it possesses no other feature calling for special remark.

232. *Pole-setting.*—Poles are, as a general rule, planted in the ground to a depth of one-fifth of their length. They should never, however, be buried less than four feet, and need not be more than six in good solid earth, no matter what their length may be. In embankments, and all made

or loose ground, they are planted about a foot deeper; whereas in rock, where blasting has to be had recourse to for the purpose of excavating the hole, they may be set a foot less than is generally the case. As a check upon this portion of the work being honestly performed, the poles, before being issued, are branded at a distance of ten feet from the bottom with a distinguishing mark, and beneath this is given the year in which they were felled. Poles planted upon a curve should invariably be set a trifle 'against their work;' that is to say, they should bear slightly against the lateral strain of the wires. For it will generally be found that when the ground has set perfectly hard the tension of the wires will have pulled them into the perpendicular position; whereas, if this precaution be neglected, and the pole be planted perfectly straight at first, the strain of the wires is almost certain to remove it from the upright, and make, apart from anything else, anything but a sightly object of it.

Too much stress cannot be laid upon good sound punning. The earth, as it is thrown in, should be thoroughly well punned at every stage: the hole should not be hastily filled up, but ample time be given to the punners to do their share of the work. Stones, if available, may be employed with advantage to assist in ramming the pole against the side of the hole where the earth has not been disturbed. Upon the punning and ramming of the holes being carried out as they ought to be depend to a large extent the stability and good working of the line when once erected.

233. The number and length of the poles employed will vary according to the route and the number of wires which they are eventually intended to carry. No hard and fast line can be drawn. For minor road lines, or the branch lines upon railways, 18 to the mile may be adopted in the straight: but on trunk lines the number should never be below 24 to the mile.

Their length will depend, not merely on the ultimate

number of wires to be supported, but on the obstacles which have to be surmounted as well. On roads 22 feet is the minimum length, except on one-wire extensions, where 20 feet may be employed. On railways 20 feet is the usual length, although on the branch lines 18 feet, and even 16 feet, have been occasionally made use of. One foot is then allowed in addition to these lengths for every two wires that have to be erected. The lowest wire should never be less than 12 feet from the ground; and at all crossings, whether on roads, railways, or of any other description, the minimum is raised to 20 feet. When it becomes necessary to vary the length of the poles, the variation should take place gradually: the appearance of the line is thereby not interfered with, and the increased vertical thrust which would otherwise be thrown upon the insulators is avoided. For instance, if in a line of 22-feet poles the necessity arises for employing a 26-feet, the pole on each side of it should be a 24-feet.

234. Upon roads and railways poles should be planted upon that side where the prevailing winds would tend to blow them off the roadway or rails. Similarly, if the route is tortuous, the inside of the curve should be selected, so that the wires may be kept as clear as possible of the traffic. Due regard should at the same time be had to the facilities for staying or strutting; and for this reason, as well as to prevent the possibility of vehicles coming into contact with them, they should be planted as close as possible to the fences on roads, and as far as possible from the metals on railways; retaining them, however, within sight of passing trains, to allow for the observation of breakdowns. On embankments and cuttings they should be placed just so far down as will admit of their being stayed both ways, and in such a position that in the event of their falling they may fall on the embankment and clear of the traffic; they are then protected also from the violence of the winds. In the case of steep cuttings the top is to be preferred to the slope, and the poles when so placed should be stayed on

both sides, as shown in fig. 116. This applies to poles in any exposed position, no matter in what direction the lateral

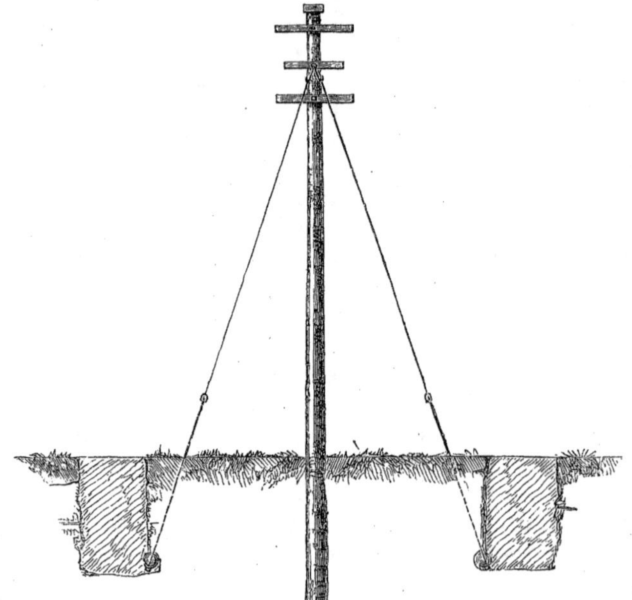

FIG. 116.

strain of the wires may be; for the influence of the wind upon the area exposed to it, and more especially when the wires are coated with snow, must be carefully guarded against in every direction.

235. Although it is very desirable to preserve the poles as nearly as possible in a straight line, yet it is highly objectionable to do so when to attain this object they will have to swing either across or over the roads. Every crossing of a road by the wires introduces an element of danger, and should be had recourse to only when absolutely essential: more than one accident has arisen from their breaking or

running back at these points in gales, frosts, or snow-storms. Occasions may of course arise when by crossing the road a decided advantage is gained; as, for instance, when by so doing the inside of a curve is for some distance secured, and less danger results from taking this step than by leaving the wires to follow the outside of the curve.

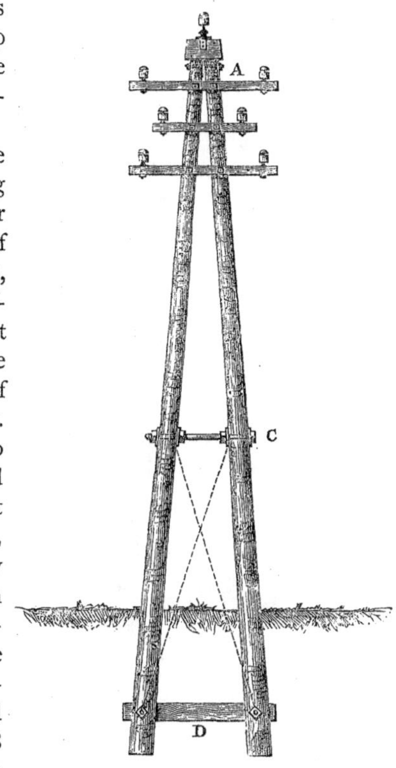

Fig. 117.

236. At points where no facilities for staying or strutting exist, or where, on account of the number of wires, sound timber of sufficient strength cannot be obtained, A poles are made use of. One of these is shown in fig. 117. It consists of two ordinary poles scarfed at the top so as to fit into each other closely, and united together by means of a bolt, shown at A. The distance between them at the base varies according to circumstances, but should never be less than 18 inches. Rather more than half way down, at C, another bolt is inserted to aid in holding them together, whilst at a distance of about 18 inches from the butts a piece of timber, D, is morticed and bolted on to both. Without this there is a tendeney for one pole to cant the

other out of the ground, which the superincumbent earth over D prevents. When the poles are long, or have to bear an exceptionally heavy strain, they are further strengthened by being braced together by two iron rods, indicated in the figure by dotted lines.

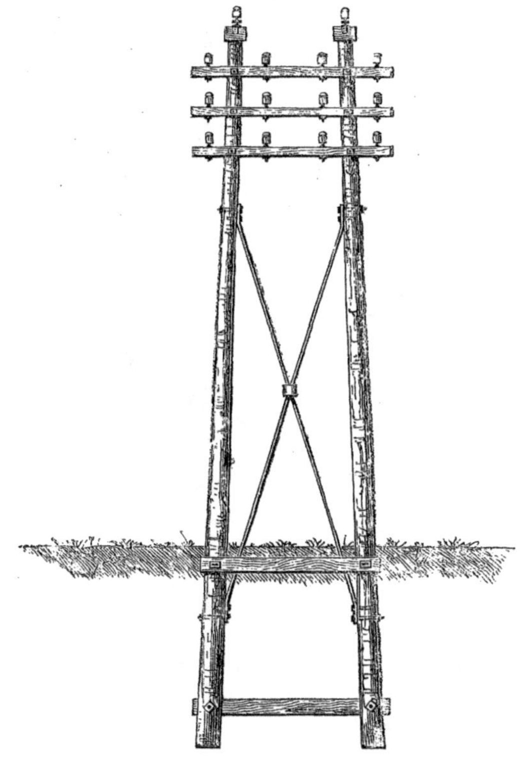

FIG. 118.

237. Where several lines converge, and the number of wires for the same line of poles thus becomes unusually great, *double poles* fitted together in the manner shown in fig. 118

are employed. Two ordinary poles are braced together by cross-pieces of wood bolted on to them on reverse sides. They are further strengthened by two iron rods placed diagonally in the manner shown in the figure, and securely attached by bolts to each pole.

238. *Tarring.*—Poles erected in their natural condition, without having been subjected to any preservative process, should be allowed to remain until well seasoned, when the ground should be opened out around them to the depth of a foot. They should then be tarred to a height of three feet above the ground line, and such upon roads as by any possibility could be run against ought to be painted white for three feet or more above that, so as to render them clearly visible. Above this they may be painted or tarred, according to circumstances. Tarring is to be preferred, unless there are local objections to its being done. The receipt for tar has been already given (§ 186); the following is the mixture for paint usually adopted in England :—

For 100 lbs. of paint—

White lead	70 lbs.
Driers	5 lbs.
Umber	5 oz.
Boiled oil	10 quarts.
Turpentine	5 pints.

239. *Numbering.*—Upon every telegraph line exceeding a mile in length the poles should be numbered after the line has been erected. The work of maintenance will be thereby greatly facilitated, for no difficulty then exists for the inspecting officer to indicate the position upon the line of whatever requires seeing to.

240. *Staying and Strutting.*—It has been already remarked (§ 232) that the stability and efficient working of a line depends in a great measure upon the manner in which the punning is done; yet occasions will frequently arise when, no matter how well this is carried out, poles cannot be made

sufficiently strong or stable to resist unaided the forces which are brought to bear against them. Artificial means must then be had recourse to, in order to supply the additional strength required; and for this purpose *stays* and *struts* are employed. By a *stay* is meant whatever takes the *pull* or *tension* of the forces acting upon the pole; by a *strut* is understood whatever takes the *thrust* or *pressure* of the forces acting upon it. The former consists of an iron wire, rope, or rod; the latter, in England, is usually timber of the same class, and subjected to the same treatment, as the pole which it is intended to succour.

241. *Stays.*—The wire rope forming these stays is sometimes supplied specially manufactured for the purpose, but more frequently and generally they are made of wire upon the spot, in which case No. 8 gauge is that which is employed. Several lengths of it--their number depending upon the work which the stay is required to perform—are laid together by hand in long lays. Twisting should never be had recourse to, nor should the wires be laid together separately; for under either of these conditions each single wire cannot take its proper strain, and the total strength of the stay is thereby impaired. No definite rules can be laid down as to the number of wires which should be used in the formation of the stay, seeing that so much depends upon the angle which when fixed it will make with the pole; yet stays of less than three wires laid together should never be employed upon roads, and this number should be increased according to the number of wires on the pole, the curve on which the pole is placed, and the angle which the stay makes with it. On straight roads it may generally be said that for a line of six wires a strand of three No. 8's will be sufficient, for ten wires five No. 8's, and for thirteen wires seven No. 8's.

242. The main object to be kept in view in the formation and fixing of the stay is to obtain the maximum amount of strength out of the materials which are employed in it.

For this purpose it should be fixed at, or as nearly as possible at, that point where the whole force which it is intended to counteract may be supposed to be collected—what is known in mechanics as the *resultant* point—and it should be placed in such a position as to form with the pole as great an angle as possible up to 90°. The resultant point may, near enough for all practical purposes, be accepted to be about midway between the top and bottom wires. The best possible position in which the stay can be placed is at right angles with the pole; as it falls from this and gradually approximates the line of the pole itself, it loses its power of resisting the tension of the wires, and to make up for this, increased strength of material becomes necessary.

243. The fixing of the stay to the pole should be carried out so that in the event of new wires being run it could be shifted with ease to meet the variation in the resultant point, and at the same time care must be taken that the pole is not cut or in any other way weakened by its attachment. For this purpose galvanised iron loops, with eyes at each end, are well adapted; the eyes being securely affixed one to each end of the bolt of the middle wire, and the stay itself to the middle of the loop. No incision is thus made in the pole, and the loop can be shifted at will as new wires are run, but as this depends entirely for its resistance upon the section of the bolt fixing the cross arm to the pole, it is not adapted for lines carrying a large number of wires. The proper method of affixing the stay to a pole carrying over six or

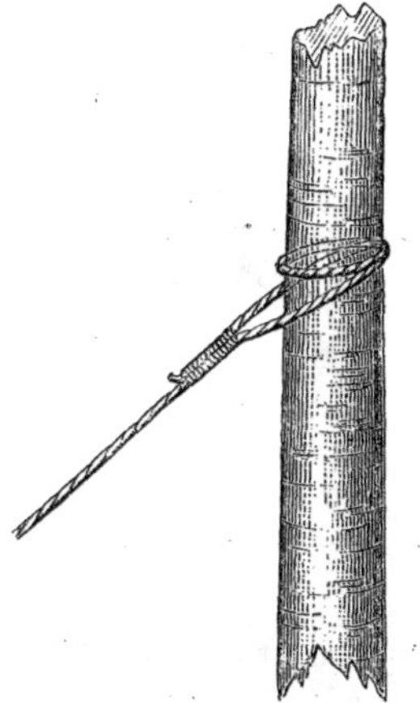

FIG. 119.

seven wires is shown in fig. 119. The wire rope takes two turns round the pole, the end of it is opened out and carefully lapped round as shown with No. 16 wire. The lower end of the stay is fixed to the eye of a galvanised iron rod (fig. 120) eight feet in length. This stay rod is passed through a block of creosoted timber three feet in length, buried to a

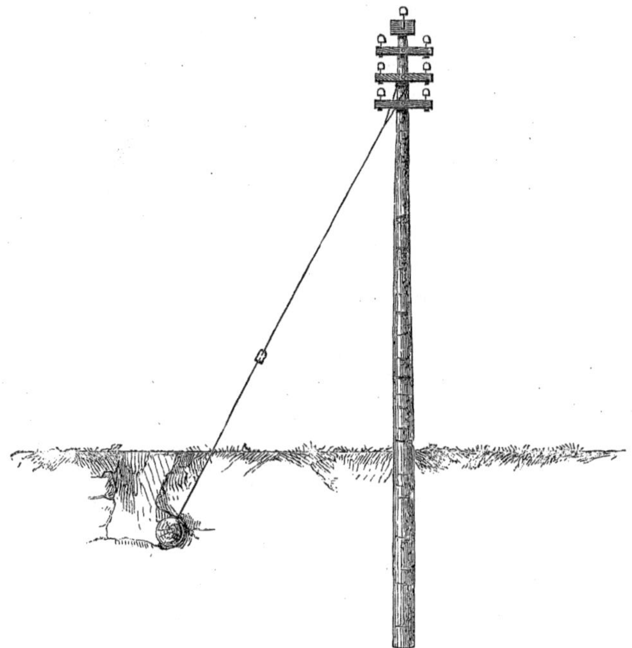

Fig. 120.

depth of from three feet six inches to four feet in the ground, and is then securely fastened to this by means of a nut and washer. The hole for the stay-block should be under-cut in the manner shown in fig. 120, so that the stay-block may have firm solid earth to press against, and thus be prevented from drawing. Where the stay passes through the eye of the stay-rod, as well as where it is fixed to the loop, galvanised iron thimbles ought to be employed. Round these the strand composing the stay is tightly turned, and its ends securely lapped down by means of No. 16 binding wire.

244. Should any difficulty exist in the way of fixing the stay to the resultant point, a forked stay similar to that

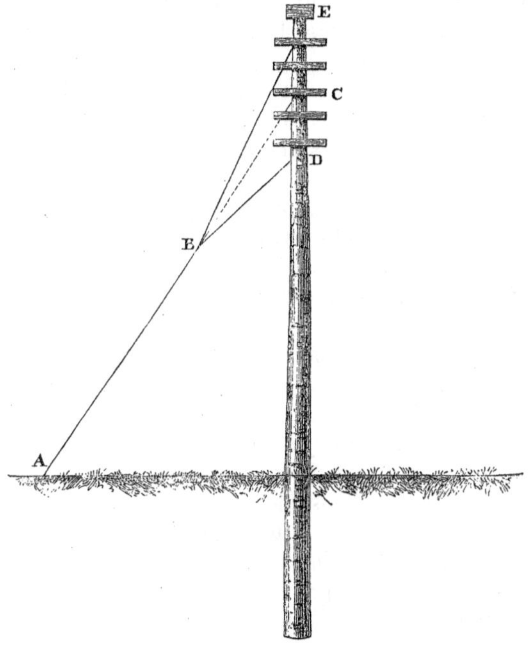

FIG. 121.

shown in fig. 121 should be employed, whose wires coming from E and D and uniting at the fork B, are continued on, and fixed to the stay-rod. In such cases (fig. 121) the two forks should be so placed that the entire stay, if continued in a straight line, would strike the resultant point.

245. Wire stays, after having been erected for some

time, lose in many instances their rigidity, unless there is a continual strain keeping them tightly drawn. Some form of tightener then becomes necessary; and although the stay-rod itself, on account of the manner in which it is fixed, answers to some extent the purpose, yet no reliance in this respect can be placed upon it. Experience so far has shown that union screw tighteners are by far the best, and they ought therefore to be employed in all cases where a danger exists of the stay-wires becoming slack.

246. Where a single stay does not suffice, or where it becomes inconvenient to remove the loop and alter the position of an existing stay, a second stay may be employed. If this is done both stays should meet at the same point, instead of having their rods fixed separately. They then come to play the part of a forked stay.

247. Numerous faults arise from the wires expanding and touching the stays, by means of which the current finds 'Earth.' For this reason the stay-wires should be at least three inches distant from the line wire nearest to them, and where this cannot be effected by applying the ordinary means of affixing the stay to the pole an iron arm or bracket, should invariably be employed.

248. Upon a line carrying a very large number of wires, it is very advisable to stay the poles on both sides in the line of the wires at a distance of about every quarter of a mile. The object of this is to prevent the poles from being driven from the upright in the event of an accident occurring to the line. The breakage of the wires either through a pole being knocked over, or from fire, imparts a sudden strain, which, unless it be resisted, makes itself felt for a long way upon the poles on both sides of the accident.

249. The greatest care must be taken in staying all terminal poles, for they form as it were the keystones of the line, and upon their being properly seen to its appearance to a great extent depends. To guard as far as possible against their yielding, even to the slightest extent, iron rods should

invariably be employed to stay poles of this character carrying more than four wires. The thickness of the rod will depend upon the number of wires on the pole; but not less than a one-inch rod should ever be made use of, and this size will be found sufficient for six wires. Beyond this an additional quarter of an inch should be added for every three wires that are erected. The stay-blocks of the iron rods employed for the terminal poles should be much larger than those of the ordinary wire-stays; they should be buried in the ground to a depth of from six to eight feet, and the ramming and punning carried out with even more than extra care. Where it is possible to attach the rod to a good sound permanent building instead of using a stay-block at all, it is advisable to do so. Upon terminal poles where the wires form an angle of 90°, or nearly so, it is preferable to place two stays, one in the line of each component strain, rather than a single stay in the direction of the resultant of these; for, by adopting the former course, provision is made against accident—in the same way as staying a crowded line in the line of the wires—from any sudden strain being thrown from either quarter upon the pole.

250. *Struts.*—Struts, from their tendency to decay, should, as a general rule, be avoided, except where the proper facilities for efficient staying do not exist. The only exception to this is in the case of soft or newly made ground, where the stay would be apt to draw, and where the strut is in consequence to be preferred. In fixing the strut the same object is to be kept in view as in fixing the stay, but one great disadvantage in using the former is the inconvenience of moving it to suit the variation in the resultant point as new wires come to be run. For this reason it is advisable to fix the strut at the outset at that point in the pole which, allowing for the future requirements of the line, would ultimately be the resultant point. The guiding principle in the erection of struts is to fix them so that they will act both as struts and stays, and

thus be able to withstand both pressure and pull. The proper method of fixing a strut is shown in fig. 122.

It is placed in the ground to a depth of not less than four feet, and attached to a creosoted block *b* (similar to the stay-

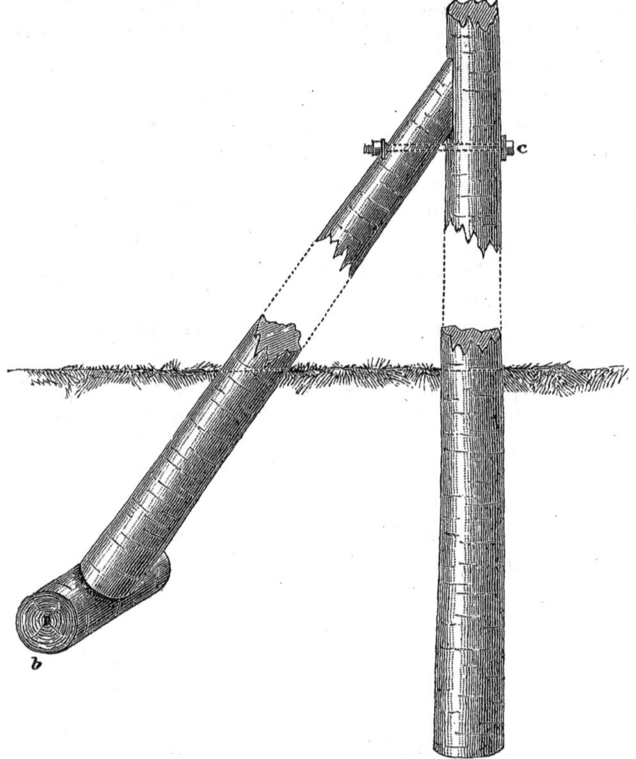

FIG. 122.

block) by means of an iron hoop and the ordinary clout nails, as shown in fig. 123; it should, like the stay, form as great an angle as possible with the pole, for the same principle regulates the position of both. The pole should not be weakened by being cut in any way, but the top of the strut should be neatly scarfed, so as to fit it as closely as possible. At the points of contact both should be carefully tarred or painted, for the purpose of making the joint water-tight. The pole and strut are firmly secured together by means of a bolt *c*, placed as shown in fig. 122. In the case of long poles a connecting tie-rod A (fig. 123) is bolted through about half-way down. As an extra precaution against the bottom of the pole being 'levered' out of the ground by the forces acting round A as a fulcrum, a cross-piece of good stout timber, B, is bolted on to both the strut and pole about 18 inches below the surface of the ground. It is needless to add that all stays and struts should be fixed before the wires are erected or any force whatever brought to bear upon the poles.

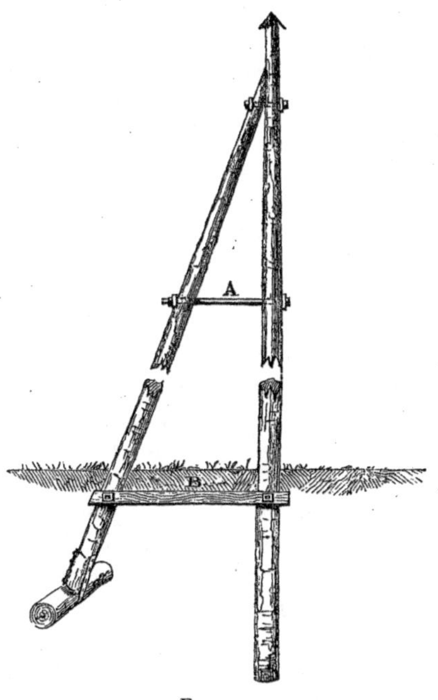

FIG. 123.

251. *Fitting up the Pole.*—This is always done before the pole is planted in the ground. The first point in fitting up the pole is to protect the top from the effects of the weather. For this purpose galvanised iron roofs, of the shape shown on the pole in figs. 124 and 127, and of a uniform size, are univers

ally employed in England. The pole is cut to fit them, and they are then nailed on with two $1\frac{1}{2}$ in. clout nails. Before the roof is nailed on, the top of the pole should be either painted or tarred. If a wire is to be run along the top of the pole, brackets of the form shown in fig. 124, and named saddle-brackets, or simply saddles, are also used. They are placed over the roofs, and 3 in. galvanised iron nails are then employed to secure both. A small aperture about an inch square is previously cut in the middle of the roof; through this the insulator bolt and nut are placed, and then screwed into a hole about an inch in depth, which has also previously been made in the top of the pole.

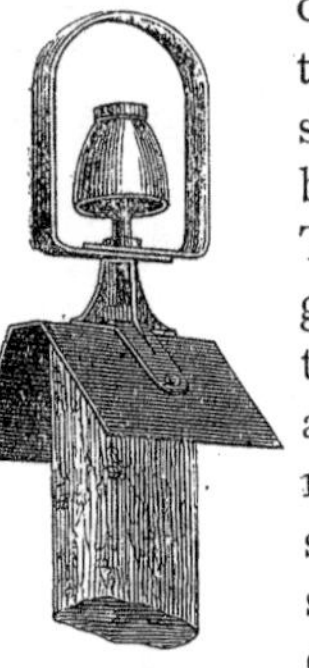

FIG. 124.

252. The supports for the insulators are either wooden *arms* or iron *brackets*, the latter being used only under exceptional circumstances. The arms in England are formed of oak, and ought to be thoroughly well seasoned previous to being issued. Two dimensions are employed, 24 in. and 33 in. in length, the scantling of both being the same, viz. $2\frac{1}{2}$ inches square. Upon double poles, and under exceptional circumstances, longer arms are made use of, measuring in some cases as much as 54 in. in length. The unequal lengths are adopted for the purpose of allowing one wire to fall clear of that beneath it in the event of the insulator supporting it being broken, or the binding giving way. They are therefore fixed alternately, the longer arm generally being uppermost. The first arm is placed eight inches from the top of the pole, and the others should be twelve inches apart, measured from centre to centre; they should all be on the same side; in England the 'up'[1] side of the pole is adopted, and

[1] The 'up' side of a pole is that which faces the 'up' station of the circuit, so that if the back be towards the up station, say London, all the arms on the pole should be facing one.

the groove into which they are fitted should never exceed $1\frac{1}{2}$ inches in depth. In the groove they are held by means of a galvanized iron bolt, which passes right through both the pole and the arm, and varies in length from $7\frac{1}{2}$ in. upwards, according to the scantling of the timber. The head of the bolt is secured to the back of the pole by means of a small washer, whilst a longer washer and nut keep it fast in the front of the arm. Figs. 125 and 126 show how these should be placed, the former giving a view of the 'up' side, and the latter of the 'down' side of the pole.

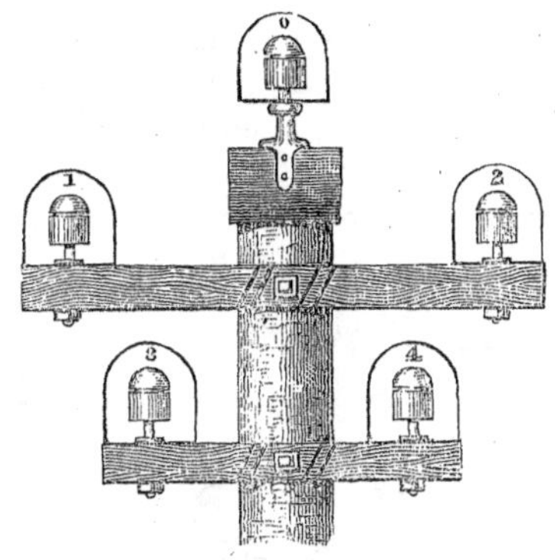

FIG. 125. Up side of pole.

253. Pole-brackets, except the saddle-brackets already alluded to, are of a tubular form (fig. 127), and made of malleable iron. They are secured to the pole by means of three 3 in. nails. They are used when a second wire has to be run along a line already carrying one wire, and where there is but little likelihood of another being required for a long time to come; they are used again on poles where brackets may have been already employed, and where it is desirable to preserve a uniform

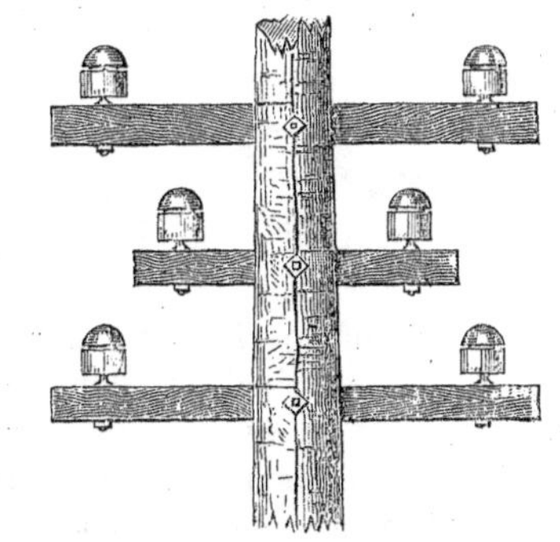

FIG. 126. Down side of pole.

appearance. They should then be placed alternately on opposite sides of the pole, and six inches apart, the uppermost one being eight inches from the top, as shown in fig. 127. They ought never to be fixed in the same horizontal plane, for if this is done the risk of contact in the event of the insulators getting broken or proving faulty is incurred. The nails would often touch each other in the head of the pole, and then, if the insulators break, form a short circuit across from one wire to the other.

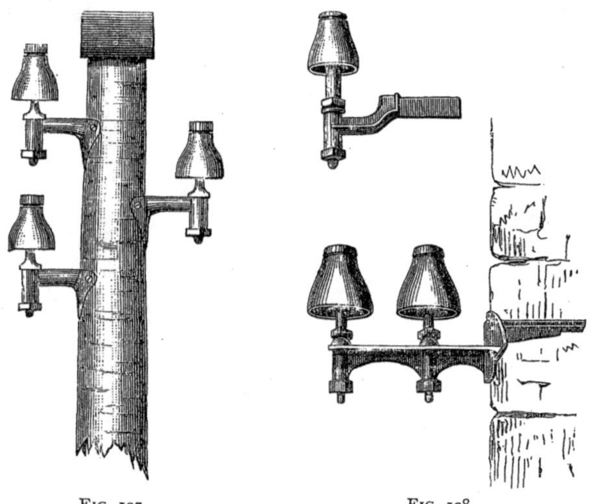

FIG. 127. FIG. 128.

Brackets of special construction, and known under the general name of single or double bridge brackets (fig. 128) are made use of when brickwork or masonry has to be employed as the support; these require no special description. The single bridge bracket is shown above, and the double bridge bracket below.

254. *Pole Raising.*—The pole is then raised and the *insulators* are next fixed to the supports, whether arms or brackets, by having the bolts screwed into the holes prepared for them,

and secured underneath by a nut; it is essential that this should be made as tight as possible. The insulators, before being actually fixed, should be thoroughly well cleared of all the dust and dirt apt to adhere to them, for these, if left, would in time seriously impair their efficiency.

255. *Shackles.*—Where the wire goes off at a sharp angle the strain thrown upon the insulator is very great, and the risk of danger through the wire flying into the road, when the outside of the curve is selected, is incurred. An insulator is not constructed to bear the heavy leverage thrown upon it when a wire is terminated. The bolt bends, and the earthenware of which it is made may break. For these reasons a special form of insulator known as a *shackle* is employed, which confines the strain of the wire to one spot, and removes the danger. It is likewise adopted when a wire is terminated, whether for the purpose of being led into an office or to avoid the possibility of danger at road-crossings. This latter is an important point, for numerous accidents have arisen through the wire having broken, and, dragging the bindings from the insulators, having dropped across the roads. At every road or railway crossing the wires should therefore be termi-

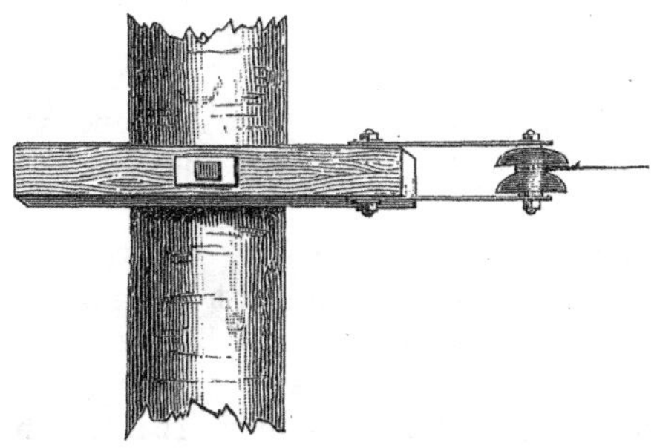

FIG. 129.

nated upon a shackle; or failing this, they should be doubly bound on several poles, and soldered as well at the last two on each side. An ebonite shackle was employed for some time; but the friction of the wires proved too much for it, and it has made way for the form known as Bright's Shackle, which, although very far from perfect, is at the present time all but universally employed, from its being the best which has hitherto been invented. Fig. 129 shows a Bright's shackle and the mode of fixing it as well. The shackle is formed of porcelain, with a hole through the centre, into which a 4½ in. bolt is inserted. Through the hole in the arm or bracket a 4½ in. bolt is placed; connecting this with the shackle-bolt are two galvanized iron straps, each measuring 8½ in. in length, which are firmly fixed by nuts to the upper and lower end of each bolt.

A *double shackle* is shown in fig. 130; the mode of procedure is precisely the same with it as with the single shackle.

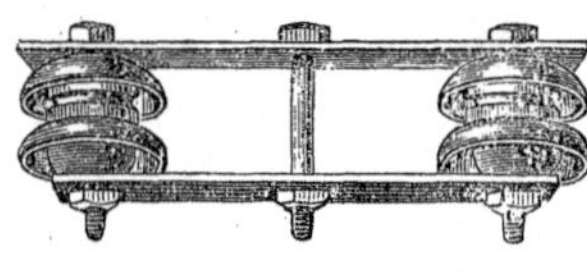

FIG. 130.

Shackles check all friction between the wire and the insulator, which, no matter how well the binding is done, must always occur to a greater or less extent when the ordinary invert is employed, and they are accordingly introduced at all points of support between which long spans are taken, and where the danger that might arise from the wire giving way is considerable.

256. *Guards.*—Upon every curve, or even upon the straight, where in the event of the insulator being broken there is a possibility of the wire coming into harm's way, guards should be employed. They are of two kinds, hoop and hook. The hoop is shown upon the saddle in fig. 124, and the hook form is shown in fig. 131. The former are now employed sparingly, especially on lines subject in winter to heavy snowstorms; for the snow, adhering to the hoop, in time brings both wire and arm into contact with each other, and, when it begins to melt, leads to a deterioration in the in-

sulation of the line. The hook guard is on this account preferable, and serves the purpose for which it is intended equally well. The guards should really be fixed before the insulator is screwed up in the manner shown; and it is needless to observe that every care must be exercised in making them as tight as possible, so as to prevent their coming by any possibility into actual contact with the wire.

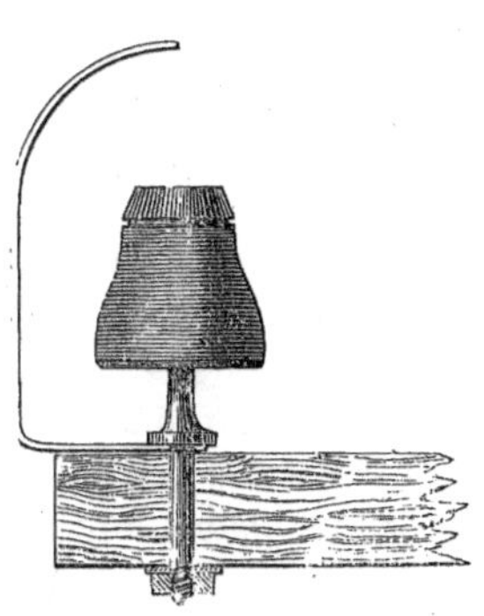

FIG. 131.

257. *Earth-wiring.*—In describing the arms and method of fixing them no mention was made of the earth-wires with which they are all fitted previous to being issued. The subject is one of such importance as to deserve special mention by itself. The object of the earth-wires is to prevent contact from arising through the leakage of currents from one wire at its point of support into another. If an insulator becomes faulty a portion of the current passing along the wire attached to it escapes; and, provided there is no other wire upon the line, makes its way entirely to earth by means of the pole. The only evil resulting from this will be a weakening of the signals, which, until the defect is made good, can be remedied by increased battery power. But if there are two or more wires upon the line the leakage from any one will then, instead of going to earth, enter that which is nearest to it—not entirely, but to an extent depending upon the electrical resistance which the pole offers in comparison with the materials intervening between the two wires. The working of both wires is thereby interfered with. An increase of battery power, instead of doing any good now, is positively injurious, for it serves merely to increase the leakage, and thereby the mischief of the contact. The only

way to get rid of the inconvenience which is caused is to afford the leakage a path to earth, whose resistance is inappreciably small compared with that which exists between the two wires. This path is afforded by the earth-wire. That upon the arm is the ordinary No. 16 galvanized iron wire, the same as is employed for binding purposes; a small groove is sawn transversely in the arm, into which one end is stapled; the wire then takes two turns around the arm, and the free end, after the arm is fixed to the pole, is attached and soldered to the main earth-wire. In fig. 125 earth wires fixed transversely on the arms are to be seen. Special care should be taken that the oak arms are thoroughly well seasoned before any attempt is made to fit them with earth-wires. Should this precaution be neglected, and the wires be fixed when the arms are in a green state, the natural acids of the oak will speedily attack them; corrosion rapidly ensues, and the earth-wires by the time the arms come to be used may be all but entirely eaten away, and are then worse than useless.

258. The main earth-wire is No. 8, and passes from the roof to the butt of the pole; a groove is also cut for its reception, and a sufficient length is left to admit of a spiral or two being formed below the pole when it is planted, so as to ensure good contact with the ground. The main earth wire is shown in fig. 126 passing to earth on the down side of the pole: it should, however, be always fixed on that side of the pole where there is least likelihood of its being tampered with.

259. It is of the utmost importance that the earth-wire should make good earth; if this cannot be secured it is better not to fix one at all, for it would merely tend to promote rather than prevent contact amongst the wires. In dry sandy soil or in rock the earth-wiring is therefore to be avoided; but if these exist to any considerable extent upon a line, it may often under the circumstances be found advisable to carry a wire along the poles specially for

the earth-wires to some spot where a good earth can be found.

260. Upon long lines, where wires are erected which are worked by very delicate instruments, the earth-wires render the greatest service, whether an insulator is actually faulty or not; for, seeing that up to the present time no really perfect insulator capable of withstanding the effects of wind and weather has been devised, the slight leakage which inevitably takes place at each would otherwise pass into the neighbouring wire, and the sum-total of these would on a line of considerable length tell upon the working of the circuits, more especially if delicate fast-speed instruments are employed. It has been urged as an argument against the use of the earth-wires that the inductive capacity of the line-wires is increased where they are adopted. There can be no doubt that this is the case, but no practical inconvenience has ever been found to result from it; and even if such did exist it could be but slight compared with the evil which the employment of earth-wires successfully obviates.

On iron poles earth-wires are of course unnecessary.

261. *Wiring.*—The poles having been properly fitted up, stayed or strutted, as the case may be, and raised, the running of the wire is then proceeded with. The coils as supplied from the manufacturers are mounted upon drums, one end being securely fastened to the terminal pole. Along the open roads the drum may be mounted on a small carriage, and the wire made to unwind itself as this moves along; but where the facilities for this do not exist a small handbarrow or a pole inserted through the hollow axis of the drum is employed. The wire is laid along the foot of the poles, it is then jointed up if not long enough, and after being carefully examined to see that the joints are sound, and that there are no flaws, it is lifted on to the arm in which the insulator intended to support it is fixed.

262. The wire is then stretched, and too much importance cannot possibly be attached to this portion of the construc-

tion of a telegraph line. The stretching is at first accomplished as far as possible by hand : light block and tackle are then applied to the wire, a species of vice, technically known as the 'draw-tongs' being used to grip it. By means of this the wire is drawn as tight as may be required, and a small vice with drum and ratchet attached is then employed to adjust the strain or regulate it. If it is pulled up too tight the wire breaks ; if it is left too slack it gets into contact with the others in its neighbourhood. Both extremes must be carefully guarded against. When simply placed on to the arm the wire dips or hangs in a curve. This curve diminishes and approximates more closely to a straight line the tighter the wire is drawn ; in other words, the dip or *sag* depends upon the tension of the wire. The maximum tension with which wires are drawn is one-third of their breaking strain; for instance, No. 8 and No. 11 wire, whose breaking strains are respectively 1,200 and 650 lbs., should never be drawn up with a tension greater than 400 lbs. for the former and 200 lbs. for the latter. When this tension is employed it is found that the dip will be about 24 inches in a span of 100 yards, and this is, therefore, accepted as the standard by which all wires are regulated. The dip must of course be varied according to the temperature of the air when the wire is run, and the 24 inches in 100 yards is that which is recognised, at 60° Fahrenheit; if the weather be warmer than this a greater dip must be allowed to admit of the wire contracting under the influence of the cold; if the temperature be lower than 60° Fahrenheit the wire should be pulled up tighter, so as to prevent its expanding inordinately under the influence of the heat, and thereby incurring the risk of contact.

263. The dip of any span varies directly as the square of the distance between its points of support ; and bearing this in mind, it becomes a simple matter to find out the dip for any span, preserving the tension uniform throughout the entire line. For instance, what dip should

be allowed in a span of 80 yards, 24 inches being allowed for 100 yards? The proportion becomes—

$$24 : x :: 100^2 : 80^2;$$

from which it is found that $x = 15\frac{1}{2}$ inches; and expressing this law in general terms, if s be the span whose dip x is required in inches, then $x = \frac{24 \times s^2}{100^2}$. This simple formula is of considerable practical value, as many problems start themselves in the construction of an open line of telegraph which can be very simply solved by it. As an example, the following may be taken, which is by no means of rare occurrence.

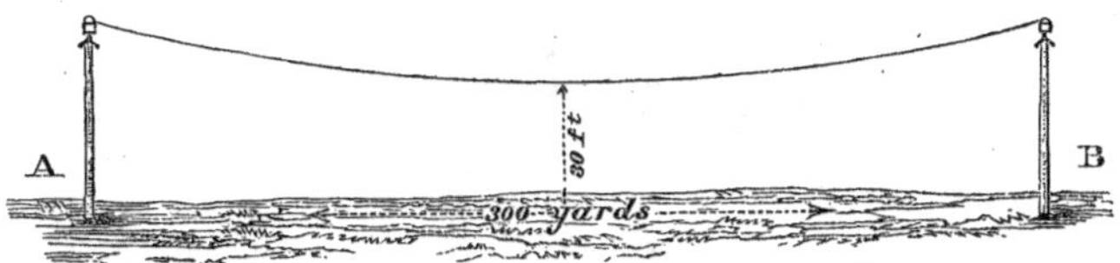

FIG. 132.

Between A and B, two points 300 yards apart, a wire is to be run without any intermediate support. What should be the height of the wire at A and B, so that at no point may it be less than 30 feet from the ground?

$$x = \frac{24 \times 300^2}{100^2} = 18 \text{ feet.}$$

Consequently the wire at A and B must be 48 feet from the ground.

264. When very long spans have necessarily to be taken a composite wire is adopted, that is to say, a wire varying in diameter from a minimum at the centre of the dip, where the tension is evidently least, to a maximum at the points of support, where the tension is the greatest; and, as already remarked (§ 198) steel wire is occasionally employed for the same purpose.

265. If one wire upon a line of poles is once properly regulated the regulation of all the succeeding wires that are run may be taken from it, and becomes a very simple matter. For, as they are all of the same metal, they will all, although of different gauges, take exactly the same dip, with the same proportional strain.

266. *Binding.*—The wire having been duly stretched, is next placed in the groove of the insulator, and very tightly bound to it, as shown in fig. 133.

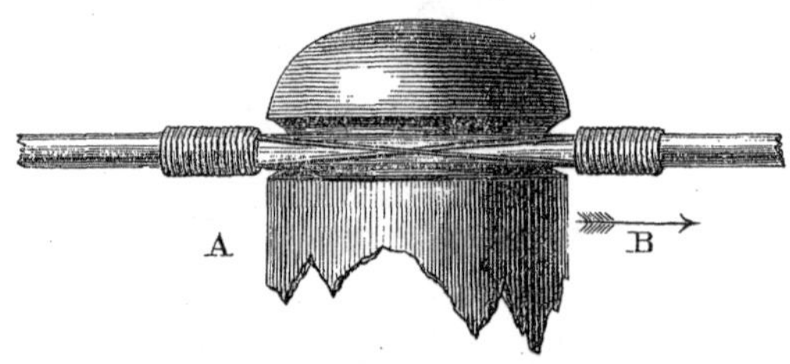

FIG. 133.

No. 16 galvanized charcoal iron wire, as has been already remarked, is universally employed for binding purposes. It is cut up into lengths varying from 36 inches for a No. 11 wire to 48 inches for a No. 4, and is made use of as follows:—It is first of all twisted around the line wire at A, then drawn as tightly as possible entirely round the insulator, and continued on at B in the direction of the arrow. It then goes back in exactly the same manner, only in the reverse direction, and is finally terminated at A. This is the binder for a No. 11 or No. 8 wire; if a larger gauge is used another lap and turn round the insulator of the binding wire should be taken. When extra strength is required the bindings are soldered.

267. If the position of a pole has, either on account of renewal or from any other cause, to be altered, care should be taken that every trace of the old binders is removed, unless, indeed—as at road-crossings—they have

been soldered on to the wires. Should any portions of them be allowed to remain they are apt, by rubbing against the wires, to wear them down to such an extent that at the first touch of frost they are broken asunder.

268. *Numbering of Wires.*—It is desirable that the wires when once erected should each have a distinguishing number, and should, if possible, occupy the same position upon each pole on the line along which they are carried. The following system, applicable to both road and railway, and independent of the side on which the poles are planted, is that which is now generally adopted:—Where a wire is run on a saddle that is invariably known as No. 0; then, standing with your back towards the up station—that is to say, looking at the up side of the pole—the wire on the left hand side of the top arm is No. 1; that on the right hand side, No. 2; the wire on the left hand side of the second arm, No. 3; that on the right hand side, No. 4; and so on. The numbering of the wires where there are two upon each arm is shown in fig. 125. Similarly, when there are four upon an arm they will be numbered as shown in fig. 134, because when the wires are transferred to short arms, 3 and 4, 7 and 8 naturally fall into their proper places.

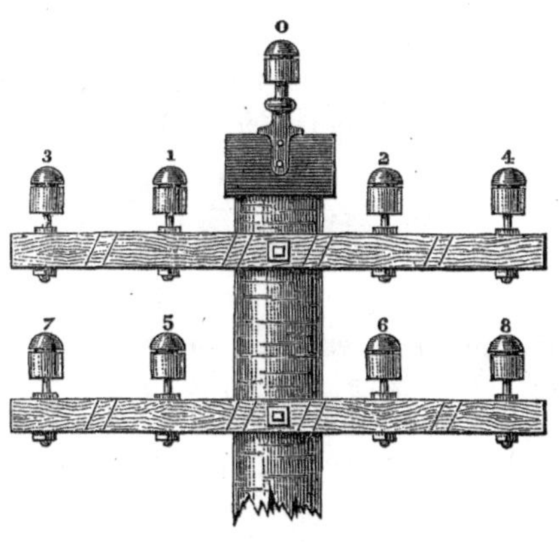

Fig. 134.

269. *Joints.*—Bad joints in telegraph wires have given rise to more trouble than any other cause, for not only have the faults caused by them been more numerous, but the

time spent in localising them has been greater than is the case with faults of any other description; for as each joint in open wires is generally at some little distance from the support, the examination of it is a tedious and difficult matter.

The first joint made in open telegraph wires was the common bellhanger's joint, which consists in merely hooking the two wires together. Binding wire was next employed, and continued over the hooks in order to ensure the continuity, which, owing to oxidation, was frequently destroyed. This joint is shown in fig. 135. The vibration to which the

FIG. 135.

wire is constantly subject when erected in the air was found to act upon the binding wire, and render the joint unsafe. A further modification was thus rendered necessary, and the bolts and nuts were added as shown in fig. 136. Upon

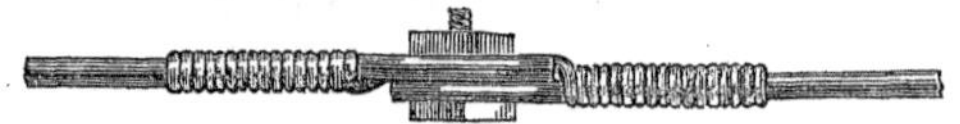

FIG. 136.

the oldest wires some of these still exist, although most of them have been cut out, for they proved to be a source of infinite trouble. The lapping wire was destroyed, and oxidation setting in between the bolt and wire, broke down the circuit.

270. The form which is now universally adopted for iron

FIG. 137.

wires is that introduced by Mr. Edwin Clark, and known as the 'Britannia' joint. It is shown in fig. 137.

The ends of the wires are carefully scraped clean and laid side by side for a distance of about two inches; they are then bound firmly together with the ordinary No. 16 binding wire; over this is smeared as a flux the chloride of zinc, formed by killing free hydrochloric acid, technically known as 'spirit of salts,' by the insertion of a few small pieces of zinc; the solder, without which no electrical joint can be considered perfect, being then applied welds the whole together in one solid metallic mass, and renders the electrical continuity complete. Any excess of the chloride of zinc should be wiped off, so as to prevent its acting upon the wire, and no two joints should ever be made close together when shackling, for the chloride of zinc, if allowed to accumulate to any extent in the space separating them, speedily eats through the wire; three inches at least should be allowed to intervene between them. The extremities of the two wires should be cut off as close as possible to the joint, so as to prevent their hooking into the neighbouring wires and causing contacts when swayed by the wind; for the same reason no joints should be made more than twelve feet from the poles; when at a greater distance than this they are apt to hitch up when the wires are blown about by strong winds. Chloride of zinc must never be employed as a flux in the case of copper wires, nor even when an iron and copper wire are soldered together; rosin is always to be used under these circumstances.

271. *Terminating.*—The wire should be invariably terminated on a shackle (§ 255). This shackle is placed on an arm, which is fixed in a line with, instead of transverse to, the wire, and the method in which the wire is attached to it has been already shown in fig. 129.

It is simply bent round the porcelain and then bound in exactly the same manner as an ordinary joint, with the exception that it need not be soldered. The present form of shackle is anything but well suited for this purpose:

the current can escape either by the upper or lower portion of it; and for this reason an effort is at present being made to introduce a form of invert in which the bolt passes nearly to the top of the insulating material, and which, to all appearances, ought to answer the purpose better for the smaller gauged wires at least.

When the wire has to be terminated, or 'shackled off,' as it is termed, at intermediate points the following is the mode of procedure which should be adopted: A double shackle is fixed, and each side is first 'tailed,' that is to say, a wire is passed round the porcelain and bound in the ordinary way, leaving one end projecting to a distance of from eighteen inches to two feet. To this end the line-wire is firmly bound and soldered, and is then bent round at a distance of not less than six inches from the pole, and similarly dealt with on the opposite side, so that the line-wire itself is continuous.

272. The leading-in wire from the terminal pole, consists of a copper conductor insulated with gutta percha, and well protected by a coating of tarred tape served around it. Tape saturated with white-lead instead of tar was tried, but proved a signal failure. This wire is bared for a distance of several inches, then wound round the iron wire and soldered only at the end, so as to admit of its being disconnected for testing purposes when required; and as gutta percha, when exposed to the effect of wind and weather, rapidly deteriorates, the wire is carefully protected either in iron piping or wooden boxing down the pole until it is led inside the office. Any small portion that may unavoidably have to be left unprotected should be well served with a second coating of tarred tape. On to the square terminal pole a hollow facing or casing is fixed, down which the leading-in wires are led; this is preferable to cutting grooves, which merely tend more or less to weaken the strength of the timber.

273. An important point to notice is, that in no case

should gutta percha be brought into contact with creosoted timber; the oil of the creosote exercises a destructive influence upon it. Care should be taken that the leading-in wires, when carried underneath the flooring, should be protected from the possible attacks of rats, which in more than one instance have been known to gnaw through the gutta percha, and having laid bare the conductors, brought them into contact with each other. The leading-in wire should likewise be kept clear of the *leaden* gas pipes; a distance of not less than six inches should intervene between them, for during a thunderstorm great risk is incurred if there is a possible line of discharge between the leading-in wire and any leaden gas-pipes in the neighbourhood. Several instances of damage have occurred owing to the lead having been fused and the gas ignited by the lightning. The same danger does not exist with an *iron* pipe.

274. *Earth.*—This, although the last point to be seen to in the construction of a telegraph line, is one of the most important, for without a good earth connection satisfactory working upon any circuit becomes an impossibility. The first object to secure is a good damp soil, and next to that as large a conducting surface as possible; for this reason a metal pump or, better still, the iron water-pipes of a town are taken advantage of, and in most instances good earth is obtained by soldering the earth-wire securely on to them. But if there are no water-pipes, and an *iron* gas-pipe is at hand, it will be found to answer the purpose; when both gas and water exist the earth-wire should be well soldered to each. Upon no account whatever is a leaden gas-pipe to be employed for the purpose of affording earth; the danger incurred by their being even near to the wires has been indicated; that danger is multiplied to a great extent when the wire is attached to them.

When neither iron water-pipes, a pump, nor iron gaspipes can be procured, a plate of metal from two to three feet square, usually of galvanized iron, is buried in the ground at a depth sufficient to ensure its being always damp, and the earth-wire is attached to that. Care must be taken that in short circuits or those where delicate instruments are employed earth is obtained from the same sort of metal at each end. Unless this is seen to, a permanent current is set up, for the two dissimilar metals being united by a conductor, the necessary conditions for a current are present. Iron water-pipes, for instance, at one end and a copper plate at the other would give rise to this, and the combination of different metals must therefore be avoided.

B. Overhouse Telegraphs.

275. In large towns, where it becomes impossible to plant poles for the support of the wires, overhouse telegraphs are had recourse to. They should be adopted, however, only when the number of wires is comparatively small; if a large number have to be run, or are likely to be required, underground work is to be preferred.

276. In the construction of overhouse lines nothing but the very best materials should be employed. The supports are iron standards, whose length will vary according to the conditions of the work. They are fixed into sockets planted upon the ridges of the houses, or placed in 'chairs.' These chairs are generally made of iron, although occasionally wood is employed. A hole is cast or bored in them as the case may be, and into it the pole is firmly fixed. All poles employed in overhouse work should be stayed in every possible direction.

277. The conductor employed is a strand usually composed of three No. 16 wires; this it is found is not only less liable to break, but causes less noise from its vibrations than a solid wire. Where exposed to the action of smoke or the gases which are given off in the neighbourhood of most of the centres of industry, it is covered with tarred hemp or tape, which serves to protect it from their destructive influences.

278. Shackles are invariably employed as the insulators to lessen the friction which is inherent to the long spans that have to be taken, as well as to reduce to a minimum the risk arising from the breakage of the wires. The thoroughfares should be crossed as far as possible at right angles, and not longitudinally; the shorter the length of wire hanging over them the less liability is there of danger occurring.

279. In soldering the joints at each point of support the utmost caution should be observed in the use of the fire-pot. Instances have occurred where from carelessness and negligence with it on the roofs of houses the leads have been melted and the building set on fire. In leading in from iron standards as well as from all iron supports, the smallest possible quantity of gutta percha wire should be left exposed, for if the insulating covering is chafed through or decays, earth is readily obtained.

280. When the standards cannot be fixed, and chimneys have necessarily to be taken advantage of instead, great care should be exercised in their selection; none but those which upon examination are proved to be perfectly sound should be tried, and even in these brackets should never be inserted, but an iron band encircling the entire chimney should be employed. A very frequent objection urged by the owners of buildings against the attachment of the wires is the noise which they cause. If the binding is imperfectly performed, or the wire strained too tightly, the vibration conducted down the solid walls proves in time to be an almost intolerable nuisance; in frosty weather, as might be expected, it becomes worse and worse as the wire contracts. Various efforts have been made to surmount this: the bolt of the shackle has been padded with chamois leather, india-rubber, and the like, the wire itself as it passes round the insulator being encased in the same material. This has been found to answer fairly; but the plan which serves the purpose best, and effectually puts a stop to the noise, is the

insertion of a small section of chain in the line-wire upon each side of the shackle. To the extremity of the chain the wire is doubly bound and soldered and the vibration is thereby destroyed. Too much care cannot be exercised by the workmen in the erection of overhouse wires.

281. The damage done to the buildings where the supports are fixed, as well as to those intervening over which the wire has to be drawn, should in every instance be rectified the moment it is observed; the dislodgement of slates and tiles, unless speedily seen to, becomes in time the source of great expense, and forms one of the main barriers in the way of overhouse telegraphs.

C. Covered Telegraph Lines.

282. Upon open lines short lengths of covered wire should be avoided as far as possible; occasionally, however, they are rendered necessary by local causes, whilst through tunnels and in towns they are decidedly to be preferred, not more for economical reasons than on the ground of safety in working. In the earliest underground lines copper was invariably employed in the conductor and has ever since been retained, but the insulating material has varied considerably, and even to the present day there is a difference of opinion as to whether gutta percha or indiarubber—the two rival substances—is to be preferred for this purpose.

283. Covered wires through railway tunnels are laid in wooden boxing, the top of which should be tied by iron wire instead of being nailed on. Where exposed to the likelihood of being interfered with by the public, screws may be used, but not nails. In driving nails danger is always more or less incurred of piercing the gutta percha, and thereby causing faults. The boxing is supported upon hooks driven into the brickwork of the tunnel. The timber employed for the purpose should be tarred, but never by any chance creosoted; creosote in contact with gutta percha exerts a marked influence upon it, and speedily leads to its

deterioration; under no circumstances should the two substances be brought together.

284. The earliest underground wires placed upon the roads in England were laid in grooved boarding formed from creosoted Baltic timber. This plan was after a time gradually discontinued, and has been entirely abandoned. In place of boarding, either cast-iron or glazed earthenware pipes are now universally employed. The employment of one of these in preference to the other will to some extent depend upon the circumstances of supply and price; but as a general rule cast-iron pipes are made use of at road crossings, or where the line which has to be followed is likely to be subjected to the pressure of heavy traffic, and may have to be opened from time to time, whilst earthenware should be employed in every other case.

The gauge of the pipes will vary according to the number of wires that are to be, or are likely to be drawn into them before their renewal becomes necessary. In no case is it advisable to lay a pipe of smaller gauge than one inch, and the following may be accepted as a general rule for guidance upon the point:—

From 1 to 8	No. 7 prepared G.P. wires		1	inch	pipe.
,, 8 to 16	,,	,,	1½	,,	,,
,, 16 to 24	,,	,,	2	,,	,,
,, 24 to 48	,,	,,	2½	,,	,,
,, 48 to 72	,,	,,	3	,,	,,
,, 72 to 128	,,	,,	4	,,	,,

The interior of the pipes, whether they are of iron or of earthenware, should be carefully scraped and cleaned before they are laid, for the purpose of removing any inequalities on the surface due to imperfect manufacture. If these are allowed to remain, the risk of injury to the gutta percha is incurred when the wires come to be pulled in. Steel dies or cylinders, rather smaller than the interior of the pipe, may be used for this purpose; or if there is any difficulty in procuring these, a heavy iron chain will be found to answer the purpose very well in iron pipes, and a rod with a split iron end, whose two sections are kept apart by a spring, will effect the same object equally well in the stoneware pipes.

285. Cast-iron pipes are generally laid to a depth of eighteen inches; in no case should the depth of the trench be less than twelve inches, and where the traffic is exceptionally heavy the limit should be to at least twenty inches. Earthenware pipes are buried to the depth of at least twenty-four inches when there is the likelihood of their being subjected to any severe pressure; in some cases, as under the pavements of towns—a position which should be invariably selected, if possible, and where the traffic, being mainly confined to foot-passengers, is comparatively light—a depth of from fifteen to eighteen inches will suffice. The joints, both in cast-iron and earthenware pipes, should be made with good Portland cement or with fine clay, so as to admit of the water percolating through. Where the pipes run side by side with gas-pipes, it is desirable to metal the joints, so as to prevent any gas that may escape from entering them. In any case, whether metal, cement, or clay is employed, a stopping of yarn should be rammed into the socket of the pipe before the joint is made, so as to exclude any particle of dirt or foreign matter from the wires, and at the same time by rendering the pipes perfectly rigid prevent their being subjected to any jar likely to cause a breakage. In filling up the trench every care should be exercised to remove all stones of any size until a depth of six inches of good mould has been punned down over the pipes.

286. As each pipe is laid in its place, an iron wire of No. 11 or No. 8 gauge is threaded through it; to the end of this the cable to be pulled in is attached. The iron wire is carried through the pipes at the time they are being laid; it is next to impossible to thread it through for any length after they are laid; the difficulty in doing so is almost incredible until it has once been experienced. At distances

of 100 yards apart, where the line is straight, and less if the route is at all tortuous, 'flush' boxes are laid to facilitate the operation of pulling in. The name flush-box

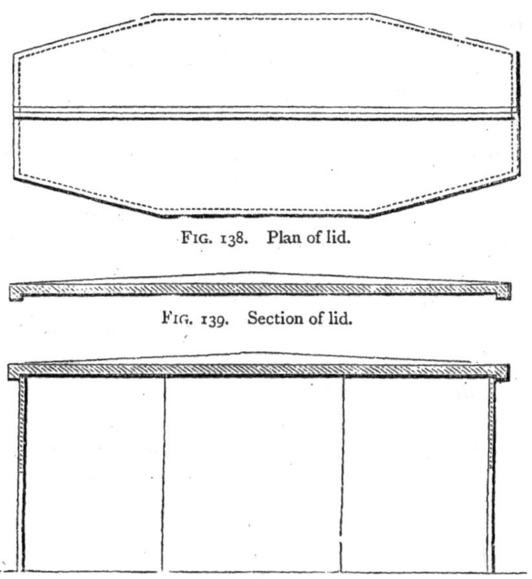

FIG. 138. Plan of lid.

FIG. 139. Section of lid.

FIG. 140. Section of box.

was originally given to these from the fact of their being laid level with the surface of the ground. As the cable to be pulled in should be manufactured in lengths of 400 yards, every fourth box of this class becomes a joint box, in which the junction with the succeeding section of cable is made. These boxes are of cast iron, measuring two feet six inches in length by eleven inches in width; they have an

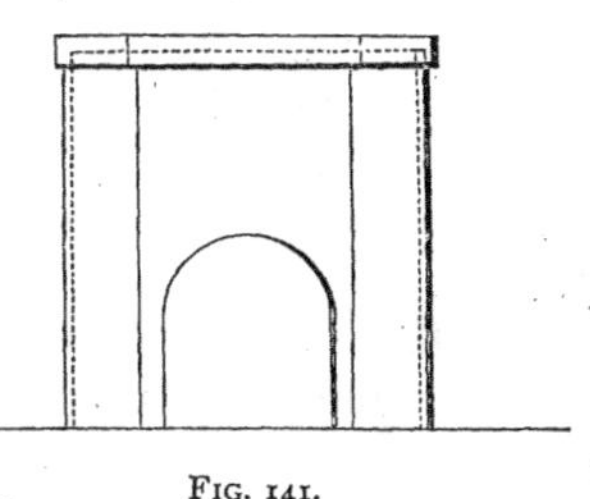

FIG. 141.

opening at each end, which varies in size according to the gauge of the pipe employed on the line. Figs. 138, 139, 140, and 141 show the construction of one of these boxes for a line of 3-inch pipes. The pipes are led into the boxes so as just to project inside them; the space around each pipe is stopped, in order to prevent the ingress of dirt. Closely fitting iron lids (figs. 138 and 139) are placed over them, and they are then covered with about a foot of earth. In order that their position may be readily ascertained a distinguishing mark should be placed on the pavement if such is available, and failing that, a wooden stake or paving-stone should be inserted to indicate the spot where they are laid.

287. As a general rule the wire employed for tunnel and underground work in England is that known as No. 7, prepared gutta percha. The copper conductor is No. 18 gauge, and is insulated with gutta percha up to the gauge of No. 7; it is then served with a covering of tape which has been well soaked in Stockholm tar. When several wires have to be drawn in at the same time, they are first of all laid side by side and tied together at short intervals forming what is technically called a 'cable'; as they are pulled into the pipes the binders are cut and removed. Occasionally the plain gutta percha wires are laid parallel to each other, and the whole are then served over with a covering of tarred tape, whilst in some instances a true cable is formed of a strand of plain gutta percha wires wound together and protected with a coating of tape steeped in Stockholm tar. The first named is the plan which is most widely adopted at present.

288. The 'cable' is placed upon a drum revolving on a good stout frame at a convenient distance from the flush-box where the work is commenced, so as to prevent its chafing as it is drawn into the pipes. To still further guard against this a wooden roller is placed at the mouth of the pipe, and a mat is spread at the bottom of the box, which

has been previously well cleaned out, so as to prevent the cable from dragging any dust or dirt along with it. The ends of the copper wires of the cable are stripped for two or three inches of their covering, and are hooked on to a loop formed in the end of the iron wire which, as already remarked, has been threaded into the pipes; the ends are lapped over with tape and yarn to prevent abrasion of the gutta percha as they are drawn through. The work

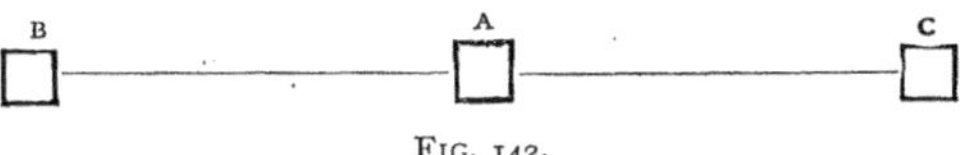

FIG. 142.

of hauling in commences in a straight length of 400 yards from the central box, thus: (fig. 142); one end of the cable is drawn from A to B, and the other from A to C. Where there are more than one intermediate box, the work of pulling in is increased with each additional box; thus in fig. 143, the half of the cable would first of all be drawn from A to *b*:

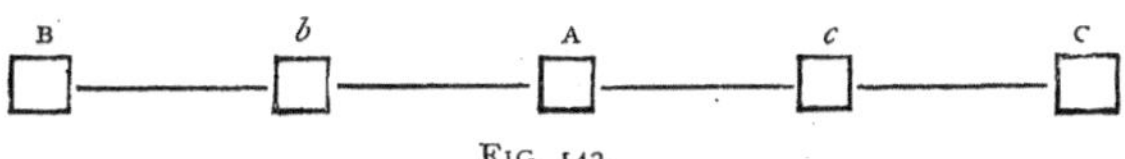

FIG. 143.

where it would be coiled and subsequently drawn in from *b* to B, while on the other side an exactly similar operation would be repeated by drawing the second half first at *c* and then at C. The drum being placed so as to give a straight lead to the cable into the pipe at the first hauling-in box, the work of pulling in is commenced; one man sees to the proper uncoiling of the cable from the drum; another attends to the lead, and the rest pull the iron wire through at the further box until the end of the cable makes its appearance there. In the case of intermediate boxes, such as at *b* and *c*, the cable drawn out of the pipes between A and these points is coiled upon canvas, care being taken to protect it from friction by means of a small roller as it emerges from the pipe. Before being pulled into the sections *b* to B and *c* to C the cable is 'turned' over by being re-coiled on to canvas on the opposite side of the boxes at *b* and *c* in order to give it a fair lead to the mouth of the pipe.

289. When the section of cable is got into the pipes the numbering of the wires is proceeded with; from a small portable battery a current is sent along each wire and noted at the further end upon a galvanometer. Corresponding numbers are then affixed to the ends of each wire in succession until all have been gone through. These numbers consist of the small leaden pellets with the numerals imprinted upon them.

290. If in hauling in the cable a wire be broken, the broken section is pulled out, and allowance being made for the resistance of the wire, the locality of the breakage is measured. Should this be but a short distance from one of the flush boxes, an iron wire is threaded through to it and a fresh section drawn in; but if the distance be too great to admit of this, a wire with its end hooked is threaded through from the point where the wire is broken and another from the flush box. By giving these a circular motion when sufficiently far into the pipes, the ends are caught in each other, and the wire being then drawn out as far as the breakage, has the broken wire attached to it.

291. In order to get at the locality of the breakage a trench has to be opened, and if the pipes are iron, one of them has to be broken; where this is unavoidable a slip joint is afterwards employed to protect it—that is to say, two half pipes whose combined diameter is larger than that of those which have been laid are placed one over and the other under the break; they are screwed together and the ends tightly packed. When earthenware pipes are used it is not necessary to break any of them; if the trench is laid open for a distance of three or four joints, the pipes may be raised, and any one of them removed without difficulty.

292. *Jointing covered Wires.*—When the section of cable has been finally laid and the numbering of the wires correctly seen to, the jointing is then proceeded with. It is impossible to lay too much stress upon the importance which attaches to the proper execution of this portion of the work. Of all the operations which are carried on in practical telegraphy there is none which requires more care and attention—none which, if in the slightest degree neglected or in any way slurred over, will prove a more fruitful source of trouble. Combined with much practice and experience, it demands a close attention to the minutest details, as well as some physical qualifications which, being wanting in many, incapacitate them for the work. Before entering into the details of out-of-door jointing of gutta percha wires, it may be well to draw attention to the main points which should be most carefully seen to. Foremost amongst these stands *cleanliness.*

Cleanliness.—A lack of this is the cause of more bad joints than anything else. Not only should the jointer's hands be scrupulously clean, but he should see that the wires to be joined are equally so, the copper being scraped bright and clean, and the insulating covering freed from tar, dirt, and grease. The materials employed by him and his tools should receive the same careful attention, every trace of dirt, dust, or rust being removed from them.

Dissimilarity in the material supplied to the jointer must be guarded against; unless the materials are exactly the same as those employed in the manufacture of the wire, a perfectly homogeneous and thoroughly reliable joint cannot be made.

The physical qualifications alluded to consist first of all of perfect health. It is a well known fact, proved by experience, that the work of even the best jointers cannot, when they are in an indifferent state of health, be relied upon. In some men, again, a greasy sweat is constantly issuing more or less from the pores of their hands, and this will of itself prevent the various coatings of the joint from firmly adhering to each other.

Patience is another virtue in gutta percha jointing, especially in the open air. The difficulty of keeping the lamp alight and in applying the requisite amount of heat, especially in rough weather, must be steadily encountered. It is better to wait, and abandon the making of a permanent joint altogether for a time, until the weather moderates, rather than run the risk of making an imperfect one.

293. The following instructions,[1] compiled as the result of a very large experience in the making and superintending of joints in gutta percha covered wire, should be most carefully attended to even to the minutest detail:—

Preparatory.—The joint-box where the joints are to be made is first opened, the jointer's box, containing his tools, placed on one side of it, and then a tent placed over the box so that the opening in the tent is opposite the jointer's box.

Attached to the box should be two low stools for the jointer and his assistant to sit on, to keep them clear from the wet pavement or damp ground.

The box should be opened and the various tools, spirit-lamps, furnace, &c., placed where most handy; the spirit-lamp for the furnace should be lighted and the soldering iron heated; the gutta percha tools should, if dirty or sticky with compound, be filed and cleaned.

Great care must be taken to keep the gutta percha sheeting perfectly clean and dry.

The wires leading in one direction are then taken out and prepared for jointing.

Cleansing Wires and Numbers.—If they are in a multiple cable, by stripping off the tape about fifteen inches back and fastening round the cable, loosening the numbers and passing them down the wire to the tape (great care must be used in passing the numbers down, for unless they are quite loose they will damage the percha); when the number is passed down to the tape it should be fixed there. When each wire has

[1] From a paper upon 'Underground Telegraphs,' read before the Society of Telegraph Engineers by Mr. G. E. Preece.

been served this way, the whole of them should be cut to exactly the same length.

The same plan should be adopted with single wires.

When the above has been done to the one side, the jointer should do the same to the other side.

The dirty work ought properly to be done by an assistant.

Cleansing Wires.—The wires at both sides must then be thoroughly cleaned with white cotton waste soaked in naphtha, until each wire is thoroughly clean, free from tar, dirt, and grease.

Cleansing Hands.—After cleaning the wires the jointer should very carefully clean his own hands, and dry them well. Naphtha will be found best for this purpose. Its disadvantage is that it has a tendency to harden the hand.

The wires are then ready for jointing.

Trimming Ends.—No. 1 wire should then be taken up on both sides (it is best to begin with the lowest number and proceed in regular order), and the gutta percha carefully trimmed off each end for about 1½ inch, care being taken that the knife does not 'nick' the copper; if this should happen, the copper must be cut off at the 'nick,' and the percha trimmed back.

Making Copper Joint.—The copper wire left bare should be scraped carefully, and then the two ends being brought together so as to overlap each other, may be held by the pliers, and first one side twisted, then the other; the entire twist should occupy about three turns each way, or ¾ inch; the surplus ends should then be cut carefully and close over, and being lightly touched with the pliers, turned in, so as not to leave an edge sticking up.

Soldering.—The twisted joints should then be soldered, care being taken to knock off superfluous solder. Great care must also be taken, when soldering a joint, that no wires be immediately under it, but that the space underneath be quite clear. Hot solder dropping on gutta percha at once heats and penetrates it.

Corresponding Numbers.—The remainder of the wires should then be jointed and soldered; great care must be exercised in jointing similar wires; the jointer should himself see that the numbers correspond, and not trust to his man giving him the numbers without himself seeing that the numbers are correct.

The gutta percha jointing may then be commenced, the second spirit-lamp having been previously lit for warming the material; care, however, must always be taken that the spirit-lamps and furnace are so placed that they cannot injure the gutta percha.

Cl. an Joint.—The ends and soldered joint should first be cleaned with naphtha.

Compound.—Then a stick of Chatterton's compound should be warmed, and a small quantity put on the copper and joint, and properly tooled over, so as to cover the joint equally. Before applying the tooling-iron it should be well wiped.

First Cover.—The ends of the gutta percha are then slightly warmed and the actual ends nipped off with the fingers. One side of the percha should be well warmed for about two inches back, and then brought forward over the joint to the opposite end with a twisting motion; the opposite end, after heating, should then be brought forward over the other part in a similar manner, as far as it will go; the percha should again be warmed and kneaded together with the finger and thumb, or tooled beforehand.

Compound.—After kneading it should be warmed over slightly with the spirit-lamp; the compound should then be heated and applied over the gutta percha, not by *dabbing*, but by putting the stick on the percha and *rolling* it along; sufficient will thus be found to adhere; the compound must be again warmed and applied a sufficient number of times to go thoroughly over the percha.

The joint should again be warmed and the compound properly tooled until it covers the joint uniformly.

Second coating Gutta Percha.—A sheet of gutta percha (the gutta percha sheeting, as supplied to jointers, should be cut into strips four inches wide, and kept carefully in a bag or case), well cleaned, should then be warmed carefully over the spirit-lamp, and, when sufficiently warmed, a piece of about one inch wide should be cut off with a pair of scissors, whose edges are moistened with the lips (the pieces cut off should be put in the mouth to assist formation of saliva); the ends should then be cut off; the joint is then warmed with the lamp, and also the piece of sheet percha that is held in the hand; the sheet, which has previously been stretched, is then applied to one end of the joint, ½in. on the old core from each end of the pull down, and being firmly pressed, is drawn along the length of the joint; the superfluous end being cut off, the joint is then turned over, and the spirit-lamp applied so that the heat warms both joint and sheeting; the sheeting is then pinched round the joint so that its sides meet above the joint; the upper part is also slightly pulled, so as to make the adhesion better; the spare sheeting is then cut off with the scissors close to the joint, a warm tool is passed over the seam so as to open it again; it is again pinched up, and by so doing forces out any air that may be in it. In pinching up the last time, one edge ought to overlap the other slightly, so that the warm tool more properly seals up the seam.

By cutting off the sheet too far from the joint the seam cannot be re-opened, and by cutting off too close no seam is left, and there is necessarily a vacant space in the second covering; this is a frequent fault, and should be avoided. With the help of the tool the ends of the coating are made to amalgamate with the old material, the joint is again warmed thoroughly, and kneaded with the thumb and forefinger, care being taken to preserve its shape and to knead evenly all round; it is then rubbed up with the moistened hand.

Compound.—The stick of compound and the joint are

again warmed, and the compound is rolled over the joint from end to end in about four places, which about covers the joint; the joint is again warmed, and the compound is worked and spread over the whole joint by means of the tooling-iron, in a uniform and even manner. The joint is again manipulated with the hand, and kneaded. It is then heated for the last time, and rubbed well with the hand, well moistened. This rubbing must be done uniformly and equally all round; it tends to solidify the joint, and gives it that highly-polished and finished appearance so characteristic of the handiwork of a good jointer.

The following notes respecting the joint and its manipulation should be carefully attended to:—

Holding Core.—The jointer and his assistant should hold the wire carefully and firmly between the thumb and forefinger at such a distance from the joint as to be beyond the influence of the heat; the percha held should always be hard; if the hand be too near the joint the man will probably be pressing the material where it has been softened by the heat, and will very probably cause damage.

Twisting Joints.—In turning the joint over for the purpose of heating or tooling, the jointer and his assistant should turn it over carefully together, so as not to put a twist in the short portion, but to distribute it over the entire length. When that operation has been done the joint should be turned back in the opposite direction, so as to bring it into its original position. As, in making a joint, this twisting has to be done very often, it is very essential that this turning over and back again should be attended to, instead of turning the joint always the same way; from inattention to this and the preceding instruction, joints have been seen with a series of twists made outside the joint.

A jointer should bear in mind that a good work is known by the unaltered state of the core outside the joint, as well as by the excellence of the joint itself.

The application of the fingers to the joint is frequently

necessary; the fingers, however, should be well moistened before touching the warm material.

Whenever a joint has been touched by the moistened finger the joint should *always* be warmed with the spirit-lamp, as this drives away any moisture. *This is very important.*

If in warm weather the hand should perspire, it ought to be dried. Naphtha will do this best, especially as its rapid evaporation produces coldness.

The length of a joint should be about *six inches.*

294. The main faults in gutta percha joints from which all beginners suffer more or less, and which nothing but experience and careful attention to the foregoing instructions can surmount, are—

Bad twists in the copper.

Nicks in the copper, and one end of the twist left sticking up—the result of the trimming having been carelessly performed.

Indifferent soldering.

Eccentricity, i.e. the wire and joint being out of the centre of the core, due to bad kneading and tooling.

Air-holes, arising from bad closing of the sheeting.

Burning, due to carelessness in the use of the lamp.

Imperfect junctions and seams, leading to separation of the coatings. These are due sometimes to moisture remaining on the coats, but more frequently to a want of cleanliness.

295. Should it at any time be necessary to increase the number of wires in an existing line of pipes, the method to be adopted is as follows:—Let B A C (fig. 142), be a section of line with flush boxes at B, A, and C, and containing seven wires; it is desired to increase the number to eleven. A cable of eleven wires, equal in length to the distance between B and A, is first of all formed and joined on to the end B of the existing cable. The same precautions are adopted to protect the wires from

friction as in the case of a new line; the new cable is then pulled in at B, the old one being drawn out at A until the section B A is completed. To the old seven-wire cable, after it has been carefully examined, and any damage which the covering may have sustained has been repaired, four new wires are added, and the eleven-wire cable thus formed is drawn in from A to C. This operation is repeated throughout the entire line until the work is completed. In this way only one set of eleven joints, viz. that at the second box, becomes necessary; at each hauling-in box the four new wires have of course to be jointed.

296. Under no circumstance should any attempt be made to draw new wires into pipes which already contain existing wires without removing the latter. The friction which inevitably takes place between the old and the new wires leads to the abrasion of the protective covering in both and lays the foundation of innumerable faults, which may only begin to make their appearance and interfere with the working of the circuits some time after the laying of the additional wires has been completed.

CHAPTER VIII.

FAULTS.

297. THE faults to which every circuit is more or less liable are divided into three classes, viz.:—

1. *Disconnections.*
2. *Earths.*
3. *Contacts.*

Each of these is further subdivided, according as they are (a) *Total*, (b) *Partial*, or (c) *Intermittent.*

298. *Disconnections* are indicated by the total or partial cessation of the current.

a. *Total* disconnection is such as that produced by a broken wire, with its end insulated, a wire off its terminal, an open switch in an office, &c.

b. *Partial* disconnection is the result of an unsoldered or badly soldered joint, a dirty contact, a loose terminal, bad earth, &c.

c. *Intermittent* disconnection is caused by a bad joint, which moved, either by the wind, by passing objects, or by heat, makes and breaks contact irregularly; dirt or dust accumulating on the contact points will frequently produce the same effect.

299. *Earths* are indicated by an increase in the strength of the current at the sending end, and by a decrease in the strength, or the entire cessation of it, at the other end.

a. *Total* earth—or, as it is more generally termed, *dead* earth—is due to the wire resting on the damp ground, or touching a stay or metal in connection with the earth. In the case of a cable it would be caused by the conductor being in contact with the water.

b. *Partial* earth is the result of the insulators being cracked or defective; or it may be produced by the wires resting upon walls, posts, trees, or other imperfect conductors in connection with the earth.

c. *Intermittent* earth is produced by the wire touching at intervals conducting bodies in connection with the earth either by being blown against them by the wind, or expanding and dropping upon them under the influence of heat.

300. *Contacts* are caused by the currents from one wire passing into another wire.

a. *Full* contact—or, as it is sometimes termed, *metallic* contact—is that which is produced by the wires being hooked or twisted together; or by being firmly united by means of another piece of wire.

b. *Partial* contact is that which is produced by imperfect conductors being thrown across the wires, by bad earths, or by defective insulation on lines not earth-wired.

c. *Intermittent* contact is produced by the wires touching each other at intervals, and is due to a variety of causes which will be alluded to hereafter.

A. Faults in the Battery.

301. Disconnections, or apparent disconnections, in the circuit are the only faults which can be caused by the battery. Total disconnection would be evidenced by no current being obtained from it. This may be due to the battery wires being knocked off the terminals, or it may be caused by the two battery wires being in metallic contact with each other. In the latter case a 'short circuit' is formed, and no current whatever proceeds to the line. One of the cells may be empty, and this would produce the same result. In the trough form of battery this is caused by leakage, chiefly owing to the marine glue having been either imperfectly applied or not being of the required consistency. In the Leclanché it results from a fractured glass cell. If any of the cells in a battery be faulty, either from leakage or from any other cause, it should be bridged over and so cut out of circuit. This can be done by joining, by means of a wire, the plates on each side of it, as in fig. 144. A battery wire, again, may be broken either mechanically or by the chemical action of the cell. As already mentioned (§32), the free ammonia in Leclanché's battery not unfrequently eats through the wire, and in the gravity form the same danger is incurred from the formation of free sulphuric acid.

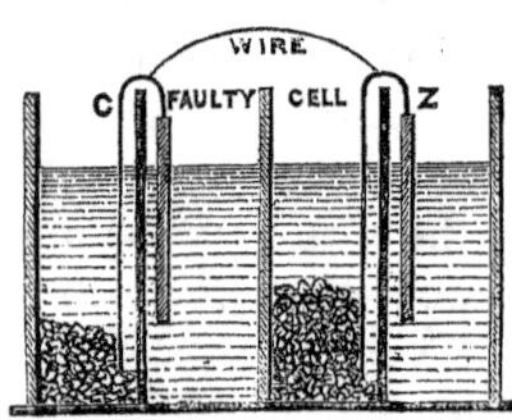

Fig. 144.

Partial disconnections are indicated by weak currents, and these are mainly due to the battery having been allowed to work too long without being attended to. The solutions having mingled too freely with each other in two-fluid batteries, or having become either too strong or too weak, as the case may be, dirty plates, cracked porous cells, corroded and dirty terminals, all militate towards the same end, and tend to diminish the strength of the current. A similar result is likewise produced by the battery finding earth (§ 27).

Intermittent disconnections in the battery—that is, intermittent currents fiom it—are usually to be attributed to the wires being loosely fixed instead of being firmly screwed down to the terminals.

302. A fault which is by no means uncommon, but which cannot be included in any of the three classes already named, is that which is known under the name of a *constant current*, i.e. a current constantly passing over the circuit. Apart from the earth currents referred to in § 141, this would be caused by either of the poles of the battery finding earth —usually from damp—at an intermediate station, whilst the other is permanently connected to the line.

B. Faults in the Instruments.

303. *The Needle.*—Faults in the needle instrument, and, in fact, in the instruments generally, may be due to either mechanical or electrical causes. The indicator on the dial of the needle instrument often remains sticking against one of the ivory pegs, the result either of damp or of the peg being partially worn away. In the former case the danger is removed by wiping the pegs with a cloth, and thus making them perfectly dry; in the latter case the pegs ought of course to be replaced by others.

Disconnections, or apparent disconnections in the needle instrument, evidenced by the weak movements of the indicator, or by its refusing to move at all, are caused by the springs of the manipulator being weakened by continued or rough working, and thus making imperfect contact; by dust accumulating on the contact points; by the demagnetization of the permanent magnet inside the ordinary coils, or loss of magnetism in the permanent magnets of the induced coils; (the carrying power of each of these should never be less than $\frac{1}{4}$ oz.; in the event of its falling below this, it ought to be removed and remagnetized.) Lightning is a fruitful source of trouble not only in this but in every form of instrument. Apart from the demagnetization or the reversal of the magnetism of the permanent magnets, the coils of every instrument are frequently fused, and special measures have to be taken, which will be described further on (§ 320), in order to guard against this.

Considerable difficulty is sometimes experienced, usually in the autumn, in working the single-needle instrument, on account of the earth currents, which then prevail with more than their usual strength. Their effect is to deflect the needle permanently; and in order to get rid, at least to some extent, of the inconvenience which is thereby caused, the dial is so constructed as to be capable of rotation. When the earth currents make their appearance the dial should be turned round, and the zero of the instrument should be taken as that point at which the needle remains deflected by them.

To prove that the single-needle instrument is in working order, it is only necessary to short circuit the instrument, by joining terminals A and B (fig. 24) together with a piece of wire, and depress the keys. Should the needle not respond, the fault will be either a failure in the batteries, a loose connection of the wires with the terminals, a bad contact between the keys and battery terminals, a broken wire in some part of the apparatus, a fused coil, or a demagnetised needle. Should the needle respond vigorously, it will show that the apparatus is in order.

304. *The Sounder.*—The Sounder, on account of the extreme simplicity of its mechanism, is less liable to faults than any of the other forms of instruments which are employed. Those which are to be met with are usually due to bad adjustment, and are the result of ignorance or inexperience on the part of those employed to work it. The

same rules are applicable to the adjustment of the direct Sounder as have been already laid down regarding that of the direct inker (§ 71). Difficulty may occasionally arise from the cores of the electro-magnet not being of thoroughly soft iron; residual magnetism then makes its appearance, and in time converts them into more or less powerful permanent magnets, thereby necessitating their removal altogether.

Disconnections are sometimes caused by the coil wires being broken at the point where they are exposed above the base-board of the instrument. This is generally caused by carelessness in dusting, but the danger is surmounted by a form of ebonite coil protector, similar to that shown in fig. 145, which, by fitting closely around the coils, leaves no part of the wire exposed. The antagonistic spring will, in the course of time, get weak and refuse to do what is required of it, but its replacement is a simple and inexpensive matter.

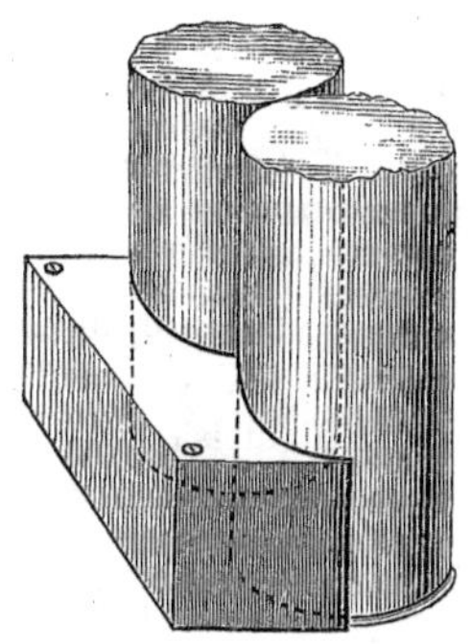

FIG. 145.

In the key the only two faults likely to arise are disconnection caused by dust or waste getting on to the contact points; a constant current, owing to the battery and line terminals being connected with each other, either by a weight pressing on the key, and the antagonistic spring being too weak to pull it on to the back stop, or from a conductor, such as a metallic pen, or the like, connecting the two parts together.

305. *The Ink-writer.*—The same faults as are to be met with in the Sounder must also be looked for in the Ink-writer, but in addition to these there are several others from which the Sounder is free.

The clockwork in the Ink-wriker is more or less liable

to become deranged; broken stop work, caused chiefly by its being over-wound, is the accident which most frequently happens. Grit again, or dust, making its way into the driving gear, will prevent the paper from running, and the friction among the various parts renders it necessary to overhaul them from time to time.

Another source of trouble inherent to the Ink-writer is to be met with in the inking arrangement; the passage between the well and the reservoir may get choked, and the disc being unsupplied with ink, no marks whatever are recorded; or, the ink becoming too thick from the accumulation of dust in the well, will render the marks altogether illegible.

No inconvenience whatever would be felt, nor any delay caused, by the defects in either the clockwork or the ink, provided only the clerk could read by sound. It therefore becomes a matter of the highest importance that every telegraphist should thoroughly master acoustic reading.

306. *The Relay.*—The only specific fault to which the relay is subject is due to the spark which passes between the points of contact every time the local circuit is completed or broken. It is the effect of the extra currents which are induced in the coils, and is strongest at the moment the circuit is broken. It is more marked in wet than in dry weather, owing to the fact that the motion of the tongue of the relay is then more sluggish; the more rapid the movements of the tongue, the less is the inconvenience felt from the spark. To prevent the metal as far as possible from being burnt away, it is tipped with platinum at the point where the spark is visible; a clean piece of paper should from time to time be gently passed between the points of contact for the purpose of removing whatever metallic dirt may have gathered there. Several methods have been adopted to prevent the spark itself from forming. The coils of the receiving apparatus are made to give much resistance. Their ends are connected with a condenser whose capacity varies with the

length of wire in the coils and the strength of the local battery. A plan, however, which has been found to answer better than this is to connect either the contact points of the relay or the ends of the coils of the receiving apparatus through a high resistance acting as a derived circuit and forming what is technically called *a shunt.* This resistance, which ought to be varied according to the strength of the local battery, should never be less than five times, and need not be more than forty times that of the receiving apparatus. The induced current will then traverse it rather than pass through the air in the form of a spark.

307. *The A B C.*—The delicate mechanism of the various portions of the A B C apparatus renders it more liable to faults of a mechanical nature than any of the instruments already alluded to. In dealing with the question of the adjustment of the A B C, reference has been made to the difficulty sometimes experienced from the endless chain in the communicator, and the steps which should be taken in order to overcome this. Other faults of a mechanical nature to be met with in the communicators are :

(*a*) Damaged teeth in the driving-wheels. This results either from a lack of oil or from the driving-gear having been taken to pieces and improperly put together again, so that the teeth do not properly fit into each other.

(*b*) The jewels into which the axle of the driving gear is fitted are frequently broken from either continued or careless working, and the armature being then jammed up against the large compound horseshoe magnet, the handle cannot be turned.

(*c*) The socket in which the axle of the armature works is sometimes insecurely fastened, or it gradually gets loosened, and then it produces the same fault as the broken jewel.

(*d*) Bad oil, becoming hard and clotted, will lead to indifferent working ; only good watch-oil should be employed in the treatment of the apparatus.

The chief complaint which is made as to the working of

the A B C is that of either 'gaining' or 'losing letters.' This, as has been already remarked (§ 84), is generally a question of defective adjustment, but it may be due more or less to one of the causes named above. Disconnections, either partial or total, are by no means rare. The former are mainly due to oxidation of the terminals or contact points; the latter are chiefly caused by the contact maker K (fig. 57) in the communicator taking up a position midway between the line and earth contact points without touching either. This is mainly to be attributed to the spiral spring L (fig. 57) which is employed for the purpose of pulling back the contact maker being too weak. The fault is an extremely troublesome one, as it may come on by any of the stations in the circuit giving a very slight motion to the handle of their communicator, and it frequently disappears without its locality being ascertained, owing to the station which caused it being in ignorance of its existence.

308. *Automatic Telegraphy.*—Only upon well-insulated lines can the full advantages of automatic telegraphy be gained. A loss of insulation is felt sooner with this than with the ordinary apparatus; it compels a reduction of speed with the automatic instruments before it is felt in general working. Still, the lowest speed of the former is always above what can be done by hand sending under the same circumstances.

The mechanical faults to which the different portions of the automatic apparatus are subject are as follows:

a. *The Perforators.*—Defective spacing is one of the main faults, and can be got over only by care and practice. Blunt punches and loose screws are to be guarded against; and care should also be taken that the paper is properly moved forward, and does not stick in any way.

b. *The Receiver.*—The paper at times runs irregularly, owing to the friction discs becoming greasy, or dust or grit interfering with their action; occasionally it sticks and ceases to run at all. The difficulties that may arise with the inking arrangement in the ink-writer—referred to in § 305—are also liable to occur in the Wheatstone receiver.

c. *The Transmitter.*—Apart from dirty contacts, which should be carefully guarded against in every form of telegraphic apparatus, but in none more carefully than this, the chief faults which are met with in the Transmitter are broken spiral springs and chains, or loose adjusting screws. The same difficulty with the running of the paper, arising also from the same cause, is experienced with the Transmitter as has already been referred to in the Receiver.

309. *Duplex Telegraphy.*—The causes of the irregularities in duplex telegraphy have been already dwelt upon (§§ 160–170) when treating of the subject generally. The duplex can be successfully worked only when the insulation of the line is constant and otherwise free from electrical defects. The smallest fault will speedily make itself felt on a duplex circuit, and in the event of earth currents, thunderstorms, or any other electrical disturbance appearing on the line, the system has for the time to be abandoned, and recourse must be had to single working. A line worked upon the duplex principle is, so to speak, subjected to a constant test, and faults which with ordinary working would probably escape observation, at once make themselves felt in duplex working.

C. Faults on the Line.

310. *Total* disconnection upon the line is the result of a broken wire. The breakage may be due to a variety of causes, but among the principal of them may be mentioned the following :—

a. A concealed weld or other flaw in the manufacture of the wire.

b. The wire having been carelessly nipped by pliers when first erected by the workmen.

c. The friction of the wire against the insulator (the result of imperfect binding in) or against a chimney or other object in its neighbourhood.

d. The friction of an old binder which has been allowed to remain on the wire.

e. The wire having been rusted away.

f. The wind, fallen trees, or boughs, travelling cranes and high loads, snowstorms, &c.

Partial and *intermittent* disconnections on the line are invariably the result of bad joints; attention has been already drawn (§ 269) to the importance which attaches to the joints being carefully seen to.

311. *Metallic* contact is the result of the wires being twisted or hooked together, or connected either by means of a short piece of wire thrown across them, or by dropping on to a metallic roof, chimney, or iron post. Apart from the ordinary causes which bring the wires together, such as the wind, high loads, workmen engaged in building operations near them, &c., a frequent source of trouble in this respect is bad regulation. This is especially the case when wires of different gauges are vertically over each other. The sun's influence upon such wires causes them to expand unequally, and so drop one upon the other; if the line runs through a cutting and is thus exposed for only a short time to the sun's rays, or if the sun becomes obscured by clouds, the wires soon return to their normal position, and the fault often disappears before the lineman can reach its locality. When the line is being erected great care should be taken to remove all pieces of short wire that may have been cut off; if any of these are left lying about, they are almost certain sooner or later to be thrown across the line wires by passers by.

Partial contact between two or more wires is caused by bodies which offer considerable resistance to the passage of electricity, such as kite-strings or cotton-waste hanging across them, or by their resting simultaneously against an imperfect conductor, such as a brick chimney or a wooden scaffolding-pole.

Partial contact not unfrequently results from bad earth,

which is often a source of trouble, especially in rocky, chalky, or sandy ground. Thus, in fig. 146, let station B

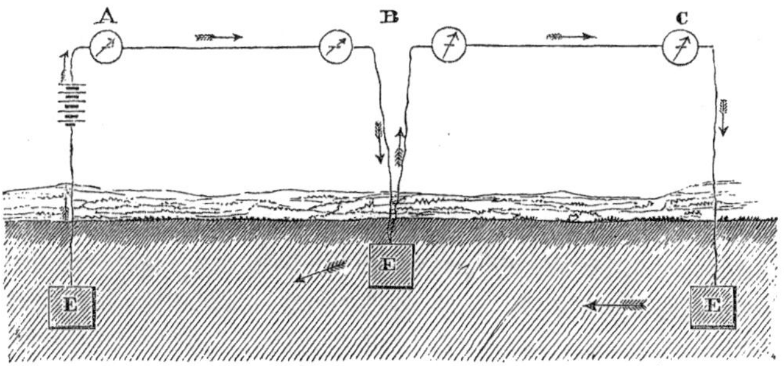

FIG. 146.

communicate with stations A and C by means of a separate circuit to each; if the earth at B is bad while that at A and at C is good, then a part of A's current, on reaching B, instead of going to earth there, will take the course of the wire to C, working C's apparatus, and go to earth at C. The effect is the same as though the wires A B and B C were actually in contact with each other, and the strength of the contact will depend upon the resistance which the earth at B offers as compared with the circuit B C. If the steps named in § 264 are not sufficient to secure a suitable earth at B, the only way of surmounting the difficulty is to run a wire from there to the nearest point where good earth can be found.

Weather contact is a form of partial contact to be met with chiefly in foggy or rainy weather, and mainly upon poles which have not been earth-wired. The leakage which takes place at the insulators there, instead of going to earth by means of the earth-wire, finds its way into the neighbouring wires, and the working of all is more or less impaired. That which runs along the saddle is least liable to be disturbed in this way, and for this reason the most important circuit is generally worked upon the saddle wire. The effect of weather contact upon the working of a circuit is very similar to that of indifferent earth ; the latter, however, makes itself felt more or less in all weathers, while the former makes its appearance only during fogs, rain, or snow.

Intermittent contacts are almost entirely due to bad regulation. The wires are swayed to and fro by the wind and brought from time to time against each other, more especially if those upon the same arm differ in gauge, and are not therefore equally influenced by it. Pieces of wire carelessly thrown across the line wires and loosely adhering to them will also give rise to intermittent contacts.

FIG. 147.

312. *Dead* earth is due to one end of a broken wire lying in water or resting upon damp ground; it may likewise be caused by the line-wire being in metallic connection with some conductor affording good earth, such for instance as an iron post, an iron stay, or the earth-wire.

Partial earth is most frequently due to broken or otherwise defective insulators; it may be also produced by the wire resting upon imperfect conductors in connection with the earth, such as walls, the guards or arms on wooden posts, trees, &c. Trees form a great barrier to the erection of a line of telegraph, and their interference is one of the main points to be guarded against in the selection of the route. When however it becomes impossible to avoid them, and when permission to lop the branches where necessary cannot well be obtained, the arrangement indicated in fig. 147 is sometimes carried out. Two poles are erected, one on each side of the road, and stayed or strutted, as may be required; between these is fixed a bar of iron supported by the arch, as shown, and into it the insulators are fixed. In this way the middle of the road, which is the part least liable to be affected by the branches, is obtained. The wires should be doubly bound and soldered at each insulator, so as to prevent their running back, and thus to reduce to a minimum the danger so likely to arise from a broken wire.

Intermittent earth is due to the wires being blown by the wind or otherwise brought from time to time into contact with some conducting body in connection with the earth.

D. Faults in Underground Wires.

313. Underground wires are free from most of the dangers to which overground wires are subject; earth currents and lightning are perhaps the only enemies which are common to both. Most of the faults which make themselves evident in underground wires, apart from those which come from the deterioration in the materials due to age, are the result of either imperfect manufacture or carelessness on the part of the workmen engaged in laying down the line. Among these may be mentioned flaws in the copper wire employed as the conductor; imperfections, such as air-holes, &c., in the insulating covering; bad joints and abrasion of the insulating covering whilst the wires were being drawn into the pipes. If reliance could be placed upon the manufacture of

covered wires, if due care were exercised upon the work of laying them, and in working them after they are laid, it is difficult to see what faults could arise until they were decayed to an extent calling for complete renewal.[1]

Rats are sometimes apt to find their way into the pipes by getting in at the bottom of the flush-boxes; they then eat through the gutta percha and either bring the wires into contact or put them to earth. Their presence may, however, be excluded by employing cement with pieces of broken glass on the bottom of each flush-box.

314. In localising a fault upon one of a number of wires lying in the same pipe, considerable difficulty is experienced in selecting from the bundle that in which the fault exists. At each flush-box the wires are numbered, and no difficulty is found there in getting hold of the proper wire; it is at intermediate points, where the wires are not numbered, that the inconvenience is felt. The old practice of 'pricking' the wires should never be had recourse to. It consisted of sticking a pin or sharp-pointed piece of metal into one wire after the other, and connecting this to a detector, on which indication would be given of the current which is kept upon the faulty wire. The holes which were thus made were either imperfectly closed up or omitted to be closed up at all, and in time developed into faults causing far more trouble than the original fault in search of which they had been made. An instrument known as the 'fault-finder' will be found to answer the purpose in picking out any wire that may be required without doing any injury whatever to it. The fault-finder consists of a pair of astatic needles hung on a curved axis, and suspended as delicately as possible, in such a way that they can be brought into a line with the wire which is for the time being under examination. Each wire is lifted into a groove between the two needles, and by their deflection under the influence of a

[1] Insulated wires should never be worked with strong currents. They cannot resist them.

constant current which is kept on the faulty wire, that which is required can be easily found. As the lengths of wire which are examined are generally under 100 yards, the current should be sent from a battery with large plates instead of from one of the ordinary kind.

315. If an underground wire becomes earthy, owing to the insulating covering being partly removed, and the conductor being thus laid bare, it should be worked with the copper current from the battery. When the copper pole of the battery is joined to the wire, a salt of the metal forming the conductor is formed by the current at the point of leakage, and this being a non-conductor, the insulation of the wire is improved. This, however, can only be done for a time, for the metal is gradually transformed into its salt, and communication is eventually broken down entirely. The action of the zinc current is the reverse of this; by depositing metallic copper its effect is to clean the wire, and thus to increase the leakage. For this reason the zinc current should invariably be used in testing covered wires, for leakages will be brought to light by it which, with the copper current, would in all likelihood escape notice.

E. Faults due to Lightning.

316. Lightning is the most fruitful source of faults upon telegraph circuits in those countries where thunderstorms are rife, and atmospheric electricity is undoubtedly the greatest enemy which those employed in their maintenance have to encounter. The damage done by it to the telegraph plant may be subdivided under two heads, viz.:

a. That affecting the poles, wires, and insulators.

b. That affecting the apparatus.

317. It is only in the case of very severe thunderstorms, when powerful lightning discharges take place, that the former is to be met with. The poles are then shattered, or have grooves cut out from the top to the ground line; the insulators are sometimes smashed, and the line wires occasionally

fused. Underground wires are free from the injurious effects of lightning, provided they are not connected to an open section of line. If, however, the latter is the case, they are liable to be affected, and numerous faults arise. Some form of lightning protector is therefore usually employed at those points where the open and covered sections are connected with each other. If the lightning finds its way into a covered wire, it will, in all likelihood, ruin the insulation at one or more points by bursting through the dielectric in its passage to the earth. The earth-wires alluded to in § 257 play the part of efficient lightning protectors to those poles which are fitted with them. Instances of earth-wired poles being affected by lightning have occurred, but the damage has never gone farther than the point at which the earth-wire commences; for this reason earth-wires should always be carried up to the roof of the pole. Upon single wire lines or loops where earth-wires are not required for the prevention of contacts, it is always advisable to earth-wire at least the last five supports on each side of every office, as a protection against the effects of lightning. In India a further precaution has lately been taken : at the last pole, before the wires enter the office, there is attached to each insulator a brass ring, into which is screwed a pointed brass rod, which is so adjusted that the point of the rod is close to the iron pole or earth-wire of the wooden pole where such is employed. Points favour electric discharge, and for this reason sparks are frequently seen passing from these points to the pole during thunderstorms.

318. Allusion has been made (§ 303) to the inconvenience occasionally met with in working the single-needle instrument owing to the presence of earth currents: beyond this no disturbance is caused by them. Lightning, on the other hand, not only interferes with the working of every class of instrument, but it frequently renders them entirely useless, and necessitates their removal altogether. Reference has been made from time to time to the demag-

netisation and reversal of the magnetism of the permanent magnets caused by it; the coil wires also are not unfrequently fused, and on rare occasions, during thunderstorms of extreme violence, the cases have been blown off and the apparatus literally shattered to pieces.

Although the dangers liable to arise from lightning have been to a great extent surmounted of recent years, still it cannot be said that a thoroughly efficient form of lightning protector for telegraphic apparatus has yet been devised. There are various forms in use at present, each of which has its special characteristic, based upon the different behaviour of electricity of high and low potential.

319. It was oberved that when two silk-covered wires were knotted or tied together, electricity of high potential was discharged across this knot in preference to going through the loop. When a discharge takes place through a non-conductor, such as dry air, at the moment of discharge the resistance along the line of discharge is so far reduced as to allow the passage of the greater part, if not the whole, of the current; so that, in point of fact, at the moment when the discharge occurs through a layer of air or other elastic medium, a conductor of very low resistance is formed. Hence, as a current divides itself in inverse ratio to the resistances opposed to it, the greater portion, if not all, flies across the knot or shunt. This is only an example of Faraday's well-known experiment, in which a long wire in air is so bent that two parts, *a b* (fig. 148), near its extremities, approach within a short distance, say a quarter of an inch. If the discharge of a Leyden jar be sent through such a wire, by far the largest portion, if not the whole, of the electricity will pass as a spark across the air at the interval separating *a* and

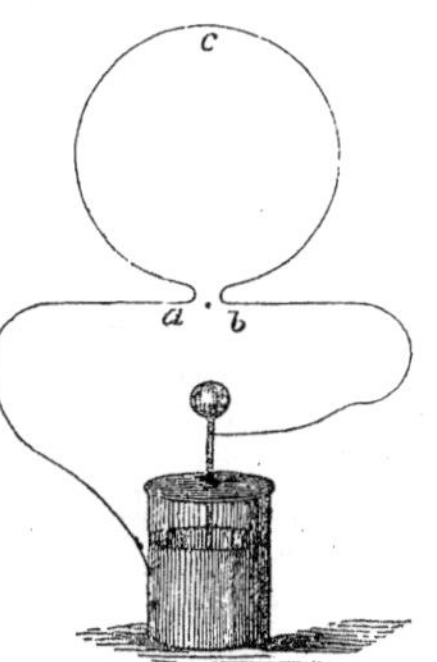

FIG. 148.

b, and not by the wire *c*. If, on the other hand, the source of electricity be a galvanic battery instead of a Leyden jar, the entire current will take the path of the wire *a c b*. Acting upon this principle, Mr. C. F. Varley, in the old form of single needle coils, simply twisted together the two ends of the coil wire before they were attached to their proper terminals, and it was found that this acted as a protector, the charge flying across the short interval in the twist, in preference to going through the coil. In order to make this system more mechanical and general, two wires of different colours were laid together and placed around the barrel of a small boxwood reel and applied to the ends of the coil. The wires themselves were still further protected by being drawn through melted paraffin, as damp affected them so as to cause contact. These protectors only act at their own expense, for though they may save the coils from the first discharge they are themselves destroyed.

320. This 'reel' protector, as it is named, was adopted for a considerable time in both the needle and Morse instruments employed in England: it was partially abandoned in favour of that known as S. A. Varley's carbon or 'lightning bridge' protector. In this form of protector the wires of the coil are attached to the termination of two insulated pieces of brass fitted in a boxwood cover, the opposed ends of which are pointed and inserted $\frac{1}{20}$ of an inch apart, in a chamber filled with a mixture of carbon and non-conducting matter in the shape of a fine powder. This protector acts as a shunt,[1] which offers such a resistance compared with that of the coil, as to prevent any appreciable diminution in the strength of the working currents; but it is assumed that in

[1] Mr. S. A. Varley, in a paper upon the subject read before the British Association, says:—'Practically no electricity would pass from a fifty-cell Daniell's battery through loose powdered black-lead or wood charcoal; but a current of 200 or 300 cells would arrange the conducting particles by electric attraction and freely pass over; while a current of 600 cells would pass across a considerable interval of the ordinary dust met with in rooms, consisting chiefly of silica, alumina, and more or less carbonaceous and earthy matter.'

the case of the discharges of atmospheric electricity the potential is so high as to overcome this resistance, and convert the bridge into a path for the discharge, in preference to the coil. They have sometimes caused inconvenience owing to the particles becoming polarized, and forming a conducting line of low resistance, and thereby practically placing the instruments on short circuit. This is remedied by simply shaking or tapping the protector, but it is certainly a defect in it.

321. Experience has further shown that this carbon-protector is not a very efficient form, and it in turn is now being replaced in England by another which has only been recently introduced. In both the 'reel' and 'carbon' protectors the lightning discharge at an intermediate station was merely 'shunted' on to the next station, and unless the protector there did its work equally well, the apparatus would be more or less affected; only at a terminal station could the discharge be put to earth direct by either of these forms. In the latest form of lightning protector the ends of the coil wires are wound side by side around a brass cylinder: this brass cylinder is connected to an earth-wire, and one of the main objections to the other forms is thus overcome. This protects the apparatus on the whole fairly well, but usually at the expense of breaking down the circuit; for immediately the insulating covering of the wire is destroyed by the discharge, the wire is fused to the brass cylinder and earth at once makes its appearance. The cylinder has then to be removed and freshly wound.

322. In India Siemens' 'Plate Discharger' is the form of lightning protector universally employed, and it is likewise used to some extent in England. It has given general satisfaction, and is, perhaps, as efficient a form of protector as has yet been introduced. It consists of two metal plates, one of which is connected with the line and the other with the earth. They are placed close together, one above the other, and are prevented from touching by the insertion of either ebonite

washers or a piece of paraffined paper. The opposing surfaces are serrated at right angles to each other, and by this arrangement discharge is promoted. Care must be taken to clean the surfaces and prevent the accumulation of dust upon them when they are kept apart by the ebonite washers. The paraffined paper, when employed to separate the surfaces, should be examined after every thunderstorm, and replaced by a new piece, provided there is any indication, usually in the shape of a hole burnt through the paper, of discharge having taken place.

323. For the protection of cables and underground work generally Varley's 'Vacuum' protector is employed, and has been found to answer the purpose very well. In this instrument an earth-wire and the line-wires are fused into a glass globe or tube which has been made a partial vacuum: this vacuum offers a ready path for the discharge of electricity of high potential. Each globe or tube ought to be tested from time to time with an induction-coil, or some other generator of electricity of high potential.

CHAPTER IX.

TESTING.

324. THE operation of examining the electrical condition of apparatus, materials, and circuits is called *testing*. The only materials employed in the construction of over-ground and under-ground telegraph lines which need be subjected to electrical tests previous to being issued are the Insulators and Covered Wire.

325. *Testing Insulators.*—This should be carried out before the separate cups are fitted together or the bolts are inserted into them. The method which is usually adopted is as follows (fig. 149):—A trough T whose inside is coated throughout with lead is filled with water, to which a little

acid is added in order to increase its conducting power. Into this trough is fitted a rack constructed so as to hold the insulating cups which are to be tested. Each of these cups C C is filled with acidulated water; around its rim is smeared a little paraffin, turpentine or grease, so as to prevent the water in the cup from coming into contact over its edge with that in the trough. In this state they are allowed to remain for at least twenty-four hours, until they are soaked as far as they can be. A powerful testing-battery, consisting of from 200 to 300 Daniell's cells or their equivalents, is employed.

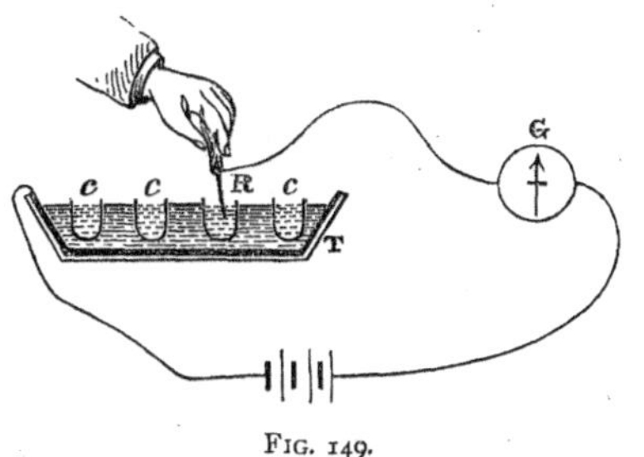

FIG. 149.

One pole of this battery is connected to one terminal of a very delicate galvanometer G—usually Thomson's reflecting galvanometer; the other pole of the battery is connected to the leaden coating of the trough. The wire from the other terminal of the galvanometer is connected to a metallic rod R, which is fitted with an insulating handle to be held in the hand. This metallic rod is then dipped into the cups in succession, and, so long as they are perfect, little or no movement of the galvanometer mirror takes place. But immediately the rod is inserted into a faulty cup the leakage which occurs through the mass of the insulator—owing to either the cup being cracked or the material of which it is composed being porous—causes the mirror to be deflected, and if the indicator passes beyond a certain limit on the scale the cup is rejected.

326. *Testing Covered Wire.*—The method adopted for this purpose is very similar to that described above for the insulators. The coil of covered wire is allowed to remain

for several hours in a tank of water, so as to ensure the water finding its way to the conductor through any defects that may exist in the insulating covering. The zinc pole of a powerful battery is then connected to one end of the coil, the other end being kept dry and clear of the water; the copper pole is generally fixed to a plate of copper which is placed in the tank. The reflecting galvanometer is inserted in circuit in the same way as was done in the previous case, and by the deflection of the mirror whatever leakage takes place through the insulating material can be ascertained. If a coil is found to be faulty it is wound upon a reel, and the wire is then drawn slowly off, and passed through the water, the connections of the battery and galvanometer remaining the same. Immediately the fault reaches the water the mirror is deflected, and the exact locality can thus be readily found.

327. *Testing Circuits.*—The subject of Testing Circuits may be conveniently divided into two parts, viz.:—

a. Testing to ascertain the condition of the circuits for the purpose of preserving a record, and to anticipate as far as possible the occurrence of faults.

b. Testing to determine the locality of a fault when its existence has once become known.

328. The former of these is carried on daily in England, and the tests which are taken are of two kinds, according as they are applied to sub-office or head-office circuits. Every sub-office on a circuit is called by the head office at the hour of commencing work, and reports the state of the signals whether 'good' or 'weak.' The head office can likewise judge of the state of the signals received from each of the sub-offices. If he fails to gain the attention of any or all of them, he concludes that there is a fault upon the circuit, and reports accordingly to the responsible officer.

Every head-office circuit is examined every morning between 7.30 and 7.45 by means of a tangent galvanometer at one of the terminal stations, while the wire is disconnected at the other end (fig. 150). The reading of the galvanometer

G is noted, and from this the condition of the wire is ascertained, and afterwards calculated out as follows :—A reading has been previously taken upon the galvanometer, through a certain known resistance, usually 1,000$^{\omega}$, and the battery power adjusted so as to deflect the needle to 45°. The object of this will be seen afterwards. The same battery power is employed in taking the readings of the wires.

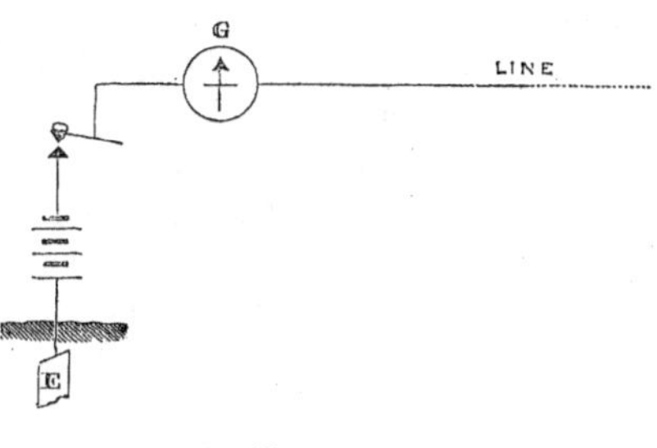

FIG. 150.

Call θ the angle of deflection obtained on one of the wires, and let it be required to find x, the insulation resistance of the circuit giving this. By the principle of the tangent galvanometer,

$$x : 1{,}000^{\omega} :: \tan 45^{\circ} : \tan \theta.$$

$$\therefore \quad x = \frac{1{,}000^{\omega}}{\tan \theta} \times \tan 45^{\circ}$$

But $\tan 45^{\circ} = 1$.

$$\therefore \quad x = \frac{1000^{\omega}}{\tan \theta}$$

Thus the total resistance of the circuit is found by dividing 1,000$^{\omega}$ by the tangent of the angle of deflection. The lower the insulation the greater will of course be the strength of the current passing into the line when the further end is disconnected, and the greater the angle of deflection of the galvanometer needle. Having ascertained the total insulation of the circuit, the insulation per mile is found by *multiplying* the total insulation by the mileage of line; for it will be evident upon reflection that supposing the leakage to be uniformly distributed throughout, and no specific fault to exist upon the line, the leakage of current upon n miles of line will be n times greater, and the insulation therefore *n* times worse than upon one mile of line. The insulation of an open wire when in proper working order should never fall below 200,000ω per mile in England, except under very exceptional circumstances, such as dense fogs, continued heavy rain, or proximity for a considerable distance to the sea.

329. It will be seen that for simplicity and rapidity in calculation the constant should be kept as nearly as possible to 45°. As the readings of wires in the normal state rarely exceed 15°, and as the constant should never be below 40°, the following table has been calculated for speedy reference. If, for instance, a wire gives 12°, the constant being 43°, 4387 will be the insulation resistance, and this, as already remarked, multiplied by the length in miles of wire tested, gives the insulation per mile in ohms.

	45°	44°	43°	42°	41°	40°
1°	57,307	55,253	53,439	51,599	49,816	48,086
2°	28,637	26,654	26,704	25,785	24,894	24,029
3°	19,084	18,429	17,796	17,183	16,590	16,013
4°	14,302	13,812	12,737	12,298	12,430	12,000
5°	11,431	11,039	10,293	10,293	9,937	9,592
6°	9,514	9,188	8,873	8,567	8,271	7,984
7°	8,144	7,883	7,612	7,350	7,097	6,850
8°	7,115	6,871	6,635	6,439	6,185	5,970
9°	6,313	6,097	5,888	5,685	5,489	5,298
10°	5,671	5,477	5,289	5,107	4,930	4,759
11°	5,144	4,968	4,797	4,632	4,472	4,317
12°	4,704	4,543	4,387	4,236	4,901	3,948
13°	4,331	4,183	4,039	3,900	3,765	3,635
14°	4,010	3,873	3,740	3,612	3,487	3,365
15°	3,732	3,612	3,488	3,368	3,252	3,139

330. A convenient form for retaining a record of these tangent tests is the following :—

Number or Name of Circuit	Section Tested		Length of Section Tested *Miles*	Galvr. reading	Reduced to Ohms	Insulation per mile
	From	To				

Upon it should be noted the state of the weather each day, the number of cells employed, and the constant, i.e. the reading obtained through the constant resistance of 1,000$^{\omega}$.

331. The *insulation* resistance is the only test which is taken by means of the tangent-galvanometer. The *wire* resistance or *conductivity* test is obtained by the differential galvanometer, or by Wheatstone's Bridge.[1] Every important wire should be accurately tested at least once a month, both for insulation and wire resistance, and the results should be carefully recorded. By comparing these with previous tests incipient faults can be readily detected and removed before they become serious enough to interfere with the ordinary working. The differential galvanometer was for some time mainly employed in taking these tests, but it has of recent years been entirely superseded by some form or other of Wheatstone's Bridge, which is in every way better adapted for the purpose. A small horizontal galvanometer is used in connection with it, and this should be as delicate and sensitive as it can possibly be made. In fact, in carrying out all electrical tests the 'detector' or galvanometer employed should be capable of giving a sensible deflection with a very minute fault, for a

[1] For full particulars as to the theory and principle of the Differential Galvanometer and Wheatstone's Bridge see Fleeming Jenkin's 'Electricity and Magnetism,' Chaps. IV. and XVI.

slight deflection upon a comparatively rough instrument is often apt to be overlooked as of little importance, although it may really impair to a great extent the working of the circuit.

332. *Wheatstone's Bridge.*—A compact and convenient form of Wheatstone's Bridge for practical use is shown in fig. 151. The thick lines are intended to represent the resistance

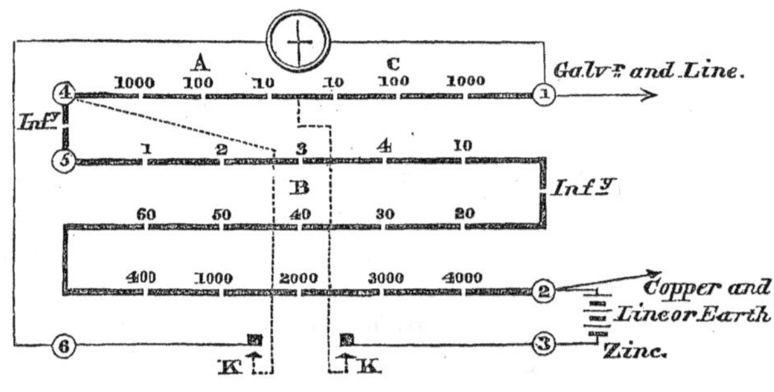

Fig. 151.

coils; the dotted lines show two connections which are not visible unless the apparatus is taken to pieces. Between terminals 1 and 4 are the arms of the bridge, each consisting of three coils offering a resistance of 10, 100, and 1,000 ohms respectively. The necessary connections are indicated upon the figure; K is the testing key by means of which the battery is introduced; K′ is another key for bringing the galvanometer into circuit. The former should always be depressed for a few moments *before* the latter; the object of this is to prevent the galvanometer from being damaged by any sudden rush of the current through the coils. The galvanometer key K′ must also make very short contacts, that is to say it should be only momentarily depressed. All that is required is an indication showing to which side the needle is deflected until the resistance coils are so adjusted that a balance is obtained. In testing with this apparatus special care should be taken that such plugs as have to be inserted are making good contact; the plugs themselves should also be kept bright and clean. From time to time the contact points of K and K′ should be cleaned by having a file or piece of emery-paper passed over them, so as to remove any dust or metallic dirt that may have accumulated upon them. The principle of this instrument is best seen from an inspection of the following diagram (fig. 152).

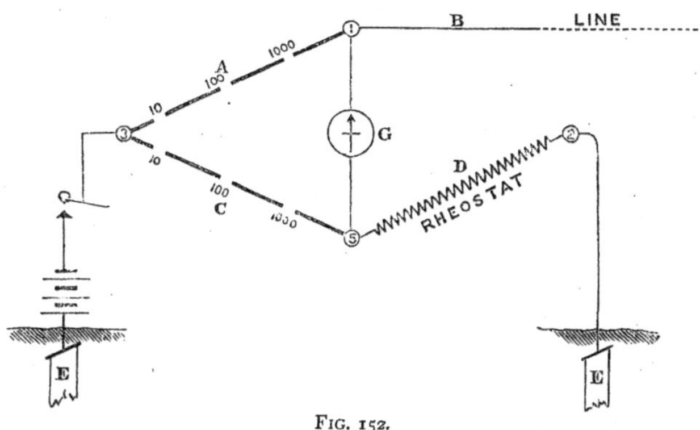

FIG. 152.

As long as the potentials of the points 1 and 5 are similar, no current can pass through G; and conversely when no current passes through G, the potentials of the points 1 and 5 must be similar. This will always happen when

A : B :: C : D

so that when the resistance of the branch A is equal to that of the branch C, the resistance of the rheostat must equal that of the line

333. *Insulation Test.*—The necessary connections required when testing are indicated upon fig. 151, and little further can be added respecting them. For the insulation test the line is attached to terminal 1, and an earth-wire is connected to terminal 2, to which the copper of the battery is likewise brought. The zinc-current is invariably employed for

insulation testing (§ 315), and a wire from the zinc pole of the battery is therefore led to the key which is in connection with terminal 3. If the test be within the total of the resistance coils, viz. 11,110$^{\omega}$, the whole of the resistance in each arm, viz. 1,110$^{\omega}$, should be unplugged. As a general rule the resistance in each arm while a wire is being tested should approximate as closely as possible to the result of the test. If the test be over 11,110$^{\omega}$ and under 111,100$^{\omega}$, then the resistance in A should be made to bear to that in C the ratio of 1 to 10, by inserting only 10$^{\omega}$ or 100$^{\omega}$ in the former, according as 100$^{\omega}$ or 1,000$^{\omega}$ are inserted in the latter. If again the test is over 111,100$^{\omega}$ and under 1,111,000$^{\omega}$, the resistance in A should bear to that in C the ratio of 1 to 100, and this can be effected by inserting 10$^{\omega}$ in the former and 1,000$^{\omega}$ in the latter. The largest resistance which can be measured by the form of bridge is 1,111,000$^{\omega}$ or 11,110$^{\omega}$ $\times \frac{1,000}{10}$, the latter factor being the highest which can be obtained from the resistances in C and A. The total insulation test being thus found, the insulation per mile is obtained by *multiplying* it by the number of miles of wire tested.

334. *Resistance Test.*—In taking the conductivity or wire-resistance test the connections are the same as in the previous case; the only difference in the arrangement is that the distant station now puts the wire to earth, instead of leaving it disconnected. The same remarks as have been made about the resistance which should be inserted in the arms of the bridge when taking the insulation test apply equally to this test. But as the wire-resistance seldom exceeds 11,110$^{\omega}$ the test obtained when 1,110$^{\omega}$ is inserted in each of the branches can be verified by varying the ratio of C to A, making it either 10 to 1 or 100 to 1, and altering the resistance coils accordingly.

335. In making the wire-resistance test it has been assumed that the distant end of the line has been put to earth, and that earth has been joined to terminal 2. Considering

the difficulty, however, which frequently exists in the way of obtaining good earth (§ 274), and the danger which is thus incurred of additional resistance being thereby inserted in the circuit, it is advisable to dispense if possible altogether with the earth, and, if a return wire is available, to make use of it. The end of this wire should then be joined on to terminal 2 in place of the earth-wire, and the distant station should be instructed to loop the two together. If the wires are of the same gauge and traverse the same route, the resistance of each will be half of the total resistance. But even supposing that they are not of the same gauge, and that a third wire is available, the resistance of each wire can then be found independently as follows:—

Let x = resistance of No. 1 wire.
y = " " 2 "
z = " " 3 "

and let

$$x + y = a$$
$$x + z = b$$
$$y + z = c$$

Then

$$x = \frac{a + b - c}{2}$$
$$y = \frac{a + c - b}{2}$$
$$z = \frac{b + c - a}{2}$$

336. The resistance test should invariably be taken with both the zinc and copper current. For although the result obtained would be the same with each, supposing the wire to be absolutely perfect throughout, yet this in actual practice is seldom if ever the case. Earth currents are always more or less present, and defective joints in the wires, as well as hidden flaws that may exist in them, introduce a disturbing element on account of the different effects produced by the zinc and copper currents at these points.

The mean of both tests should then be calculated, i.e. calling x the test obtained by the zinc current, and y that obtained by the copper current, the real test may be taken as $= \frac{x + y}{2}$.

337. *Localising faults.*—The existence of a fault having become known, the first step to be taken is to ascertain as near as possible its locality. Practically, on over-ground wires, a fault is localised by simply disconnecting or putting the wire to earth at successive stations until it is localised between two stations. At certain stations along the line the wires are led into testing-boxes for the purpose of affording facilities for crossing, disconnecting, and putting them to earth. Previous, however, to communicating with any of these offices it ought to be ascertained whether or not the fault may not be in the apparatus at the station itself. This is done by short circuiting the apparatus or, provided there be a test-box in the office, putting the wire to earth, or disconnecting it there. The testing station is always the most important station on the circuit, and is generally one of the terminals.

338. Taking first of all the case of a *disconnection.* Let (fig. 153) A D be a circuit between A and D led into testing-boxes

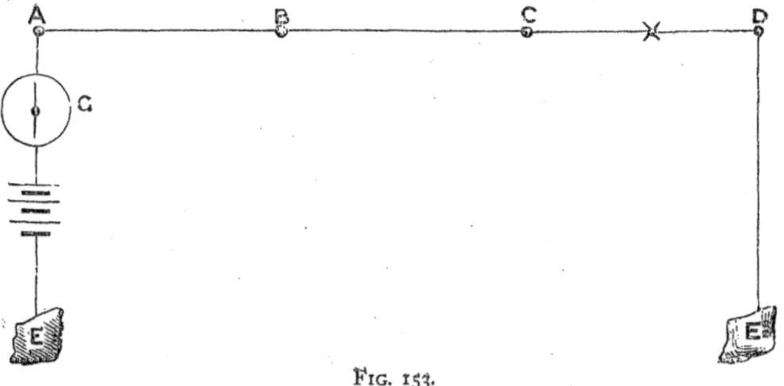

Fig. 153.

at B and C, and suppose that a disconnection has appeared upon it. Then if A is the testing station, the wire is first of all put to earth at the test-box there, and a galvanometer inserted between it and the instrument. As soon as it is ascertained that the fault is outside the office by the galvanometer being deflected, A advises B to put the wire to earth for one or two minutes. If, when this is done, the indication of the current is still obtained on the galvanometer, the fault is beyond B, and C is next advised to treat the wire in the same way, B having of course restored it at the expiration of the time named to its previous condition. If the same occur at C, D is then advised, and if the galvanometer is now unaffected (or affected but very slightly, and that simply through the normal leakage between A and the locality of the disconnection) the fault is between C and D, and the lineman is at once advised of the nature and locality of it.

With an *earth* fault a similar course is pursued, differing only in the fact that at the various test-boxes the wire is disconnected in place of being put to earth.

339. When two wires are in *contact* they are both put to earth at the testing station, and disconnected at the others.

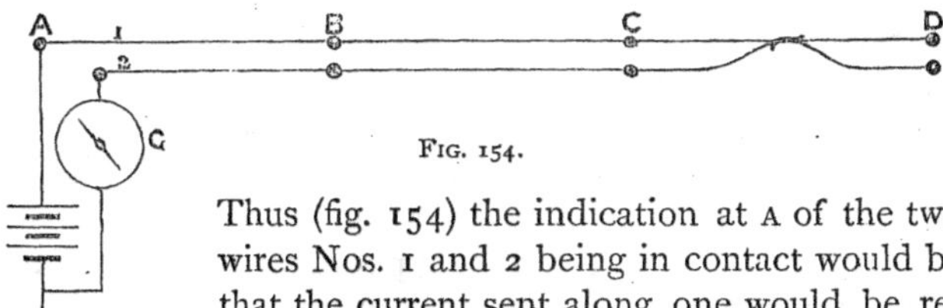

Fig. 154.

Thus (fig. 154) the indication at A of the two wires Nos. 1 and 2 being in contact would be that the current sent along one would be received on the other. To No. 1, therefore, the current is applied, and in No. 2, which is put to earth direct at the test-box, a galvanometer is inserted; B is then asked to disconnect both wires, and if when this is done no indication is observed on the galvanometer the contact is beyond B. The same is done at C. If, however, when D has disconnected, the current sent along No. 1 is received on No. 2, the fault is between C and

D. Upon no account whatever should B, C, or D put the wire to earth; no reliable test for a contact could be made if this were done, for if earth be put on near the contact the greater portion of the current would go to earth and not return to A.

340. *Crossing Wires.*—The speedy restoration of communication upon busy circuits is a matter of such importance that immediately a fault upon one is localised every effort should be made to cut it out of the circuit, and so restore communication at once. This can be done only by crossing the wire with another of less importance which may happen to be in existence between the two stations between which the fault has been localised. Suppose (fig. 155) that AE, an important through circuit between A and E, becomes faulty between C and D, and that between C and D is the

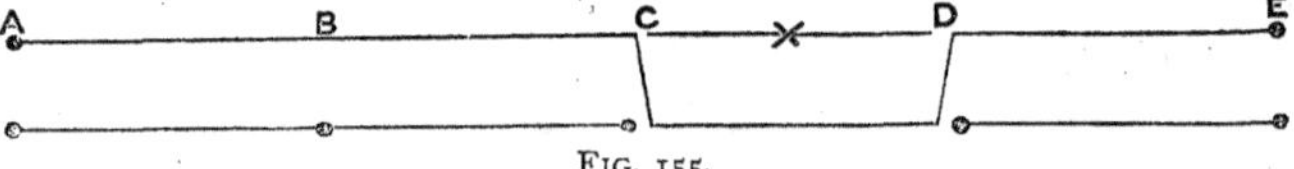

FIG. 155.

section of a less important circuit picking up the stations B, C, and D. At C and D the faulty section of the through wire is thrown out until the fault is removed. In its place is substituted the section CD of the 'pick-up' circuit. Communication is thus preserved between A and E, the former of which can transmit the work of B and C, and the latter that of D. In this way the inconvenience felt from faults is in a well-organised system reduced to a minimum, and frequently four or five wires between two important centres may have faults upon them, and yet only one of them be really broken down, provided the faults are not in the same sections. Upon trunk lines of telegraph which are traversed by important wires it becomes a question for grave consideration whether it would not be advisable to erect a spare wire for the sole purpose of restoring the normal communication as far as possible when a fault occurs upon one or more of the working circuits.

341. Every important office should have one or more alter-

native routes by means of which, in the event of its main line of communication being broken down, an outlet may be found for the traffic. Thus (fig. 156), supposing all the wires between Brighton and London are broken down, Brighton has cross country circuits to Portsmouth, Southampton, Eastbourne, Hastings, and various other towns in direct communication with London, anyone of which can, by simply removing the earth and crossing the wires in their test-box, restore communication between London and Brighton.

342. Intermittent faults are by far the most difficult to deal with; and it is often impossible, on account of their short duration at one time, to localise them by taking the steps indicated above. Where a wire subjected to an intermittent fault can be crossed with another, it is advisable to do so. Thus (fig. 155), if on the wire AE an intermittent fault

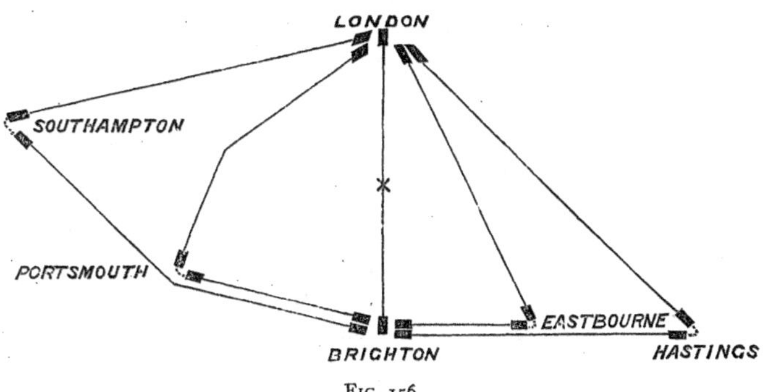

FIG. 156

made its appearance occasionally, the wire should be crossed with sections of the ABCDE wire successively between B and C, C and D, &c. and kept so until the fault disappeared from AE and showed itself on ABCDE. Only in this way can it be ascertained in what section it exists.

343. The testing at an intermediate station is exactly the same as that described for a terminal station. By putting on earth on either side, and thus ascertaining on which the fault

exists, the station does really become terminal for the time for all practical purposes.

344. The method generally adopted for ascertaining the locality of faults upon the overground lines in England is that which has been described above. The testing stations are comparatively close to each other, and a fault, having been localised to exist between two of them, can generally be removed an hour or two after the line-man has started in search of it. But upon covered wires this cannot be done, for although the fault can be localised in the same way the same facilities for examination do not exist as in the case of an over-ground line. If no other steps are taken for ascertaining the locality of a fault upon a covered wire beyond the disconnecting or putting it to earth at the testing stations, then the wire has to be cut and tested at each successive flush box (§ 286) until the defective section is found. The inconvenience and delay attending this may be overcome in many instances, by employing what is known as *the loop test*, provided another wire in a perfect condition is carried along the same line of pipes.

345. *The Loop Test.*—If the insulation of a line were perfect, a condition which is never practically attained, the localisation of faults would become a very simple matter. Thus, for instance, let (fig. 153) the wire AD find earth at a point between C and D, and suppose that the fault is a perfect earth, that is to say offers no resistance. If A is the testing station, and the wire when tested in this condition gives a resistance of 140^{ω} then allowing 14^{ω} as the resistance per mile, the distance of the fault from A is $\frac{140}{14} = 10$ miles.

But the majority of faults of this kind do offer a greater or less resistance, and the insulation of the line being more or less defective, theoretical calculations of this nature cannot be carried out in practice. The advantage of the loop test consists in its being independent, within certain limits, of the resistance of the fault. A reference to fig. 157 will show the connections which have to be made in one method of taking this test with the Wheatstone bridge. The zinc of the battery is brought to the testing key, the coppei

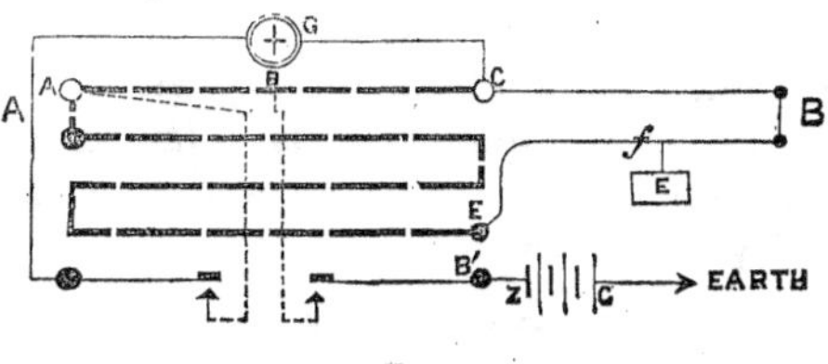

Fig. 157.

being put to earth. The galvanometer is inserted in the usual way, and the good and bad wires being joined together at the test-box of the distant station B, the end of the former is connected to terminal C, and of the latter to terminal E. The resistances in the bridge should then be adjusted until equilibrium is obtained. Then calling x the distance of the fault from terminal E, and y the distance from terminal C, according to the principle of the bridge:

$$\text{BC} : y :: \text{BA} : \text{AE} + x$$

$$\text{or} \quad \text{BC}\,(\text{AE} + x) = \text{BA} \times y \quad . \quad . \quad . \quad (1)$$

But L, the total wire resistance of the whole loop, which can be ascertained on reference to the record of periodical tests is $x + y$. $\therefore\ y = \text{L} - x$ . . . (2)

Substituting this value of y in equation (1)

$$\text{BC}\,(\text{AE} + x) = \text{BA}\,(\text{L} - x)$$

$$\therefore\ x = \frac{\text{BA} \times \text{L} - \text{BC} \times \text{AE}}{\text{BA} + \text{BC}}$$

And the values of BA, BC, AE, and L being known, the resistance of x is obtained: this divided by the resistance per mile of the wires gives the distance in miles of the fault from station A.

If the two arms of the bridge BA and BC be made equal to each other the above equation becomes—

$$x = \frac{\text{L} - \text{AE}}{2}$$

346. A second method of taking the loop test is shown by fig. 158. In this, the resistance in the arm BC should be plugged up, and BA, AE then become the two arms of the bridge. The connection being made as shown in the figure and BA, AE adjusted until equilibrium is obtained (x and y being, as before, the distances of the fault from terminals and C respectively) it follows that—

$$\text{BA} \times x = \text{AE} \times y$$

But

$$y = \text{L} - x$$

$$\therefore \text{BA} \times x = \text{AE}\,(\text{L} - x)$$

$$x = \text{L}\left(\frac{\text{AE}}{\text{AE} + \text{BA}}\right)$$

and this divided, as before, by the wire resistance per mile, gives the distance in miles of the fault from station A.

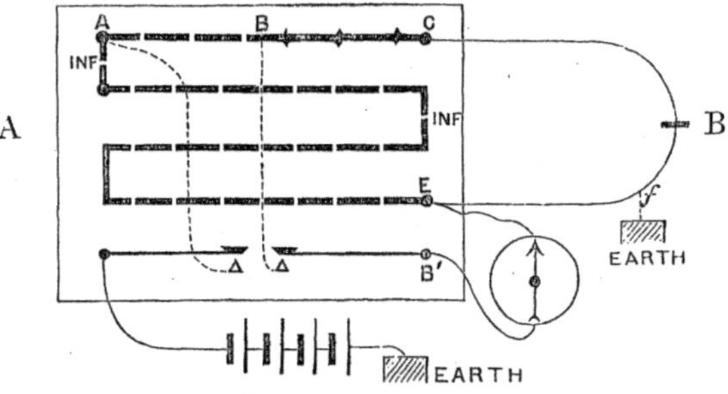

FIG. 158.

If the two wires employed have not the same resistance per mile, then the value of x must be divided by the resistance per mile of the faulty wire.

347. If two wires are in contact the distance of the contact from the testing station can be readily found, provided that the fault itself offers no resistance. The two wires form a loop whose resistance can be readily measured by means of the bridge, and half of this divided by the resist-

ance per mile of the wires, will give the distance from the testing station.

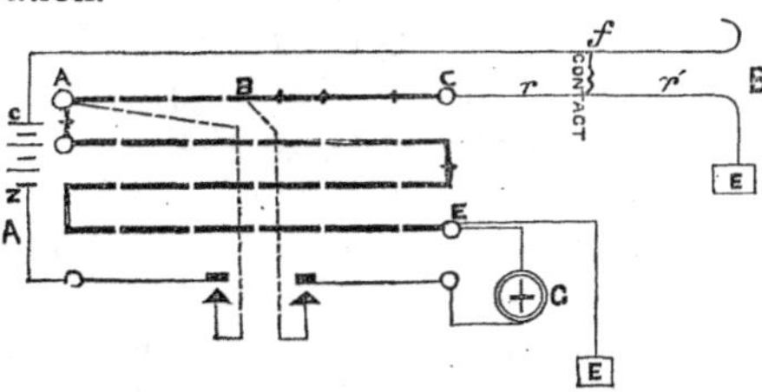

FIG. 159.

If, however, the contact offers a certain resistance, the locality can be ascertained by arranging the bridge in much the same way as was done in the second method by taking the loop test (§ 346). The connections required for this are shown by fig. 159. The resistance in BC is plugged up as before, and BA, AE thus become the arms of the bridge. One of the two wires is disconnected at station B, while the other is put to earth there. The former is connected to the copper pole, and being thus made a battery wire, does not enter into the calculation. The resistances in BA and AE being now adjusted until equilibrium is obtained, it follows that:

$$r : r' :: \text{BA} : \text{AE}$$

or $$\text{BA} \times r' = \text{AE} \times r$$

and BA, AE and $r+r'$ being known, the values of r and r' can be found, and the distance of f, the locality of the fault from station A, can thus be ascertained.

348. *Testing-Boxes.*—The manner in which a wire should be terminated has been already described in § 272, and the precautions which should be taken in leading it into the office have been likewise indicated in § 273. If the office contains but a few instruments, and is not a testing station, each wire is led direct to the instrument which it is intended to work. But if the number of instruments is considerable, each wire is led to a *test-box* and brought thence to its instrument. Test-boxes are likewise fixed at offices situated on a trunk-

line, and into them all the wires which pass are led, for the purpose of facilitating the operation of testing, as explained in § 327. No uniform plan has been followed in the fitting up of these test-boxes, but that which is generally adopted may be gathered by a reference to fig. 160.

On to a wooden frame, generally formed of mahogany, and varying in size according to the number of wires which are to be led up to it, four rows of brass terminals are symmetrically fixed. The upper and lower rows of terminals are used for the 'Up' and 'Down' line wires respectively: the two intermediate rows are 'Instrument' terminals. Between the latter a row of 'Earth' terminals is placed; the number of these varies according to the number of line-wires; but, as a general rule, for every two of the latter there

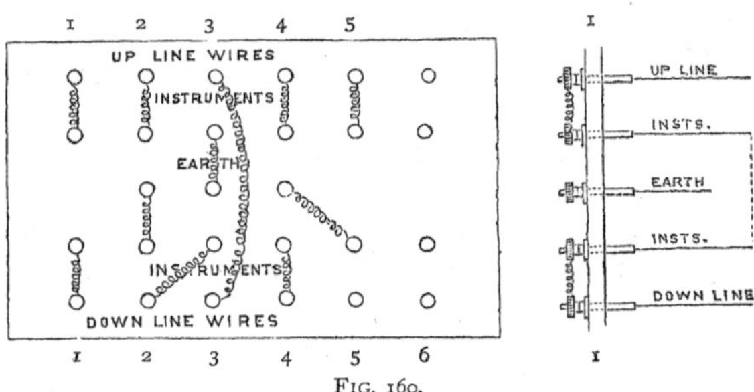

FIG. 160.

should be one earth terminal. The line terminals are numbered consecutively from left to right. Various systems have been adopted in assigning the numbers to the wires in a test-box; that which has been found to answer best is to assimilate the test-box in this respect to the terminal pole outside the office, where such exists, and so arrange the numbers upon both as to coincide with each other.

In addition to marking the numbers, it is advisable to attach bone labels to the terminals, and indicate upon these the names of the various circuits. The labels can be changed according to any alteration rendered necessary by a re-arrangement of the wires. In fig. 160 the wires going to terminals 1 and 4 have intermediate instruments joined up on them; at 2, both 'Up' and 'Down,' and 5 the necessary connections for a terminal station are shown, while to 3 the wire is brought in simply for testing purposes. It is always advisable to leave a few spare terminals, in order to provide for the normal increase of wires.

The wires are connected to the terminals at the back of the box; these connections should invariably be soldered, and, in doing so, chloride of zinc ought never to be used as the flux; rosin ought to be always employed. The earth-wires running along the back of the box should be carefully soldered to each of the terminals marked earth. The terminals themselves should be kept bright and clean, and ought always to be well screwed down, so as to prevent disconnections. To guard still further against this, the wire employed in the connections for a test-box had better be of the gauge known as No. 0 or No. 7 plain, rather than the No. 20 gutta percha covered wire which is frequently used.

CHAPTER X.

COMMERCIAL TELEGRAPHY.

349. A TEXT BOOK of Telegraphy can scarcely be complete without some explanation of the method adopted in the actual transaction of telegraphic business. The instruments in use have been described, their mode of connection together has been shown, the construction of lines has been illustrated. We have now to assume the existence of an operator at each instrument, and everything prepared for the transmission of messages.

350. *Message forms.*—There are three kinds of messages that the operator has to deal with, and each kind has its own

particular form, upon which it is written. The first or *forwarded* message is written on what is known in the British Postal Telegraph Service as the A form; it is printed on *white* paper, and it is the message which is written out and handed in across the counter by the sender himself for transmission by wire. It has to be paid for. The second is the *transmitted* message, which is written on the B form; it is printed on *buff* paper, and is the message received at a telegraph office *by wire* for retransmission *by wire* to some other telegraph office. The third is the *received* message, which is written on the C form; it is printed on *pink* paper, and is a message received by wire at a terminal office for delivery *by hand* to the person to whom it is addressed from that office.

There are other forms used for other classes of messages, which are paid for under special rates, such as *foreign* and *press messages*, or those which are sent *free*, such as messages on the service of the Post Office Department or of a railway company. Every circuit is supplied with pads of these forms, and in order that the clerk who is about to receive a message may know what particular form to use, every message is indicated by a *prefix*, which is the first signal always sent. Thus the letter 'S' shows that a C form must be used; the letter 'X' a B form; 'S G' a service form; 'S P' a press form; 'S A' a free railway message form, and so on. Moreover, these prefixes indicate the order in which these messages must be sent; for some messages, such as those on the business of the State, or on the urgent business of the Department, take priority over ordinary private messages, as these latter take priority over ordinary Service messages.

351. *Code Time.*—But while precedence among the different classes of messages is determined by the prefix, the order of transmission amongst messages of the same class and character must also be determined; otherwise, when several offices are upon the same circuit, considerable confusion would arise if each office could not determine the

priority of despatch of its own messages. This is done by the 'code time,' which is the signal sent after the prefix. Time is transmitted by code according to the following diagram:—

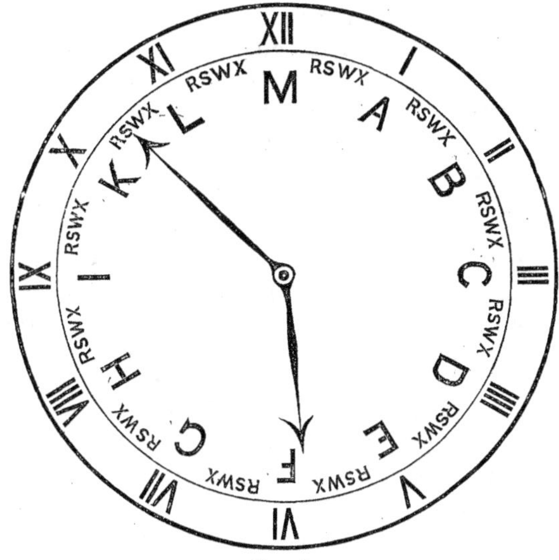

FIG. 161.

The twelve letters from A to M denote the twelve hours. They also denote the twelve periods of five minutes of which each hour is composed. The intervening four minutes are denoted by the letters R S W X. The letters sent *singly* indicate the hours, sent in combination of two they represent the hours and certain periods of five minutes; sent in connection with the intermediate letters R S W X they represent hours and minutes: Thus—

M	is	12	B	is	2	E	is	5
F	,,	6	I	,,	9	K	,,	10
M F	,,	12.30	B I	,,	2.45	E K	,,	5.50
M F S	,,	12.32	B I X	,,	2.49	E K S	,,	5.52

In England, where the despatch of messages is so rapid, and where Greenwich time is uniformly employed, there is no necessity to indicate whether the hour is a.m. or p.m.; but on the continent, where much greater distances and differences of time exist, the distinction is given by M for *matin* and S for *soir*. On the other hand the above code-time system does not exist there, time is sent in full, thus: '12.32 S' or '2.49 M' is signalled as written. The English system is far more concise and rapid.

The 'code time' is the signal always sent after the prefix, and it indicates the time at which the message was handed in by the sender. The signalling of this code time not only indicates the order of transmission of the message itself, but being signalled to the terminal station it is copied on the received form, so that the ultimate receiver of the message knows the exact minute when the message was handed in by the sender (§ 356). This information is not only very useful in itself to business men, but it forms an admirable check by which the public itself is able to supervise the working of the telegraphs.

352. *Station Calls.*—Before a message can be sent to any station the attention of that station must be attracted. It must be called, and an answer given to show that the attention of the operator at that station has been secured, and that he is ready to receive a message. In the earlier telegraphs this was done principally by bells, and the bell is still used with the A B C apparatus (§ 83); but the necessity for bells no longer exists, for each of the other kinds of instruments emits sufficient sound to attract attention. But when several stations are upon the same circuit, how is any one individual station to be selected and attracted? In the same way that in the presence of a company of men when the attention of one, say Brown, is required, 'Brown' is shouted, and Brown replies by a lusty 'here.' So every station is supplied with its name or 'code' to distinguish it from every other station, and when it is wanted it is called

by this code and to it it answers. It would be tedious and uncertain to signal the names of stations in full, hence an abbreviated name or *code* has been given to every office by appropriating to it combinations of the letters of the alphabet in groups of two and three. There are 650 combinations of *two* letters of our alphabet, and 15,600 of *three* letters, and as there are only 5,585 postal telegraph offices in England, it is evident that no 'code' consists of more than three letters.

Liverpool is indicated by	.	L V
Manchester	,, . .	M R
Bristol	,, . .	B S
Edinburgh	,, . .	E H
Maidstone	,, . .	M A
Hollingbourne	,, . .	H J X
Lenham	,, . .	L D P
Egerton	,, . .	E C N
Charing	,, . .	C E Y
Ashford	,, . .	A D

The last six stations enumerated are all on the same single-needle circuit. Every current sent on that circuit operates each instrument alike and simultaneously. Practically it may be said that they are all in telegraphic sound of each other. Suppose Maidstone has a message for Lenham: he signals on the instrument 'L D P, L D P, L D P,' Every other station hears the call and knows the station wanted, but no one answers but Lenham, whose operator, when he perceives that he is called, replies " here L D P " or simply " L D P " only. Ashford having thus secured his attention, signals the prefix and code time and proceeds with the message.

But there is an inconvenience even in the use of three letters. The combinations frequently and carelessly repeated sound so much like words that, unless the call be carefully made, an inattentive station may conceive that a message is being transmitted between two other stations instead of his own station being called. The calls are also sometimes very like each other in sound when made rapidly, and on all instruments the calls are read by sound. Again, as most sub-offices, like those enumerated above, communicate only with their own head offices, complication is unnecessary, hence, single call letters, viz., D, G, K, O, R, S, &c., for local use have been introduced. The sub-office which comes nearest to its head office is called D, the next in order G, the next K, and so on. Thus the calls on the above circuit would be—

	Code.	Call.
Maidstone . . .	M A	M A
Hollingbourne .	H J X	D
Lenham . . .	L D P	G
Egerton . . .	E C N	K
Charing . . .	C E Y	O
Ashford . . .	A D	A D

Lenham's call would, therefore, be 'G,' and he would reply 'G.'

353. *Number of words.*—When a message was handed in at the counter the counter clerk up to very recently counted the number of words in the address, which in England are signalled free, and entered that number in a column on the A form; he then counted the number of words in the body or text of the message, which alone are paid for, and made the proper entry on the form; lastly, he counted the number of words in the special instructions as to the mode of delivering the message, which were sent after the message and are signalled free, which he also entered on the form. These three figures were then summed up and the total number of words signalled to the receiving station after the code time. But now the number of words in the *text* of the message only are counted and signalled. This is done solely to secure accuracy in transmission, for every clerk immediately he has received a message counts the number of words in it, and if that agrees with the number signalled, he concludes that his

message is correctly received; whereas if the numbers disagree, he at once obtains a correction.

The *instructions* for the delivery of the message after the wire has done its work, together with the office of origin, which formerly were sent at the end of the message, are now signalled after the code time.

354. *Preamble.*—These signals, the prefix, the code time, the office of origin, the instructions, and the number of the words, form the *preamble* of the message; and, compared with the preamble in use on the continent, they are very concise. In the continental form of message the preamble consists:—

(1) The prefix.
(2) Name of station of destination.
(3) Name of sending station prefixed by the word *de.*
(4) [1]The consecutive number of the message.
(5) Number of words in message.
(6) The day of the month.
(7) The time to minutes with the indication M or S.
(8) The station transmitting when it has to be forwarded on by wire.
(9) The special instructions for delivery.

All these signals together are equivalent on the average to *ten* words, and are, therefore much more cumbrous than the English form, which, with the special instructions sent at end of the message, do not average more than six words.

355. *The forwarded message.*—After the preamble comes the address, and then the body of the message itself. These parts are separated from each other by a distinct signal, called the *break signal* (.. ..), so that the receiving operator knows on what part of the form before him he has to write down the words he reads off.

[1] On the continent every message is numbered consecutively at each office, and any enquiry about that message is referred to that number. In England we speak of a message as 'your A B Jones' by giving the code time, and name of the sender.

A

	Words.		How Paid.		
Prefix *S*	Adds.	*11*	Cash....		
	Text.	*17*			
Code *E K S*	Instrs.	*7*	Stamps.	*6*	*0*
	Total	*35*	Total	*6*	*0*

POST OFFICE TELEGRAPHS.
(Inland Telegrams.)

Sent at *6 . 0* *P* M

To *S A* By *J. Jones*

No. of Message *252*

For Postage Labels. (Stamps.)

Office Stamp.

MM. Instructions.

Man and Horse 5/- paid Northampton.

Please Write Distinctly.

Addresses Free.

FROM	TO
E. D. Teddy	*Poulet Earle Esq.*
Laura Place	*Hinton-by-George*
Northampton	*Warminster*

1/-	*The*	*Lamb*	*won*	*in*	*a*
	canter	*Master*	*Magrath*	*second*	*Pilgarlick*
	third	*and*	*the*	*Scotchman*	*last*
	as	*usual*			
1/3					
1/6					

The completion of the message itself was, until very recently, indicated by P Q; but this last signal has recently been abolished in England, because it really had become unnecessary. The ending of the message is either self-evident or it is sufficiently indicated by the prefix of the next message being sent, or by another station being called. On Morse circuits the signal ···—· is used to denote its completion. A complete message with all its entries is shown in the diagram on p. 288.

356. *The received message.*—The clerk who receives the message writes it down in pencil on the proper form which is indicated to him by the prefix. If it is an 'S' message received for delivery to the public an office copy or duplicate is kept. The C forms are made up in pads and in duplicates, thus: the first sheet is the office copy, the next the public form, and so on. The office copy, which is kept for reference, is written with a pencil, contains all the entries of prefix, code time, number of words, &c., which the public does not want to know; the rest of the message, which the public does want to know, is transcribed on the public copy by carbonic paper, which is slipped into the proper place between the two forms. Hence, the office copy is in pencil, the public copy in manifold writing. This latter form is folded up, inserted in an envelope, which is addressed and then sent out for delivery by messenger. The entries made on the C form are as follows:—

Prefix *S* Code *EKS*	POST OFFICE TELEGRAPHS. C.		Words.		No. of Message
Recd. from *T. S.*	Sent at *6 · 20 P* M.	Delivery and Charges.	Adds..	*11*	
By *T. White*	To *Hinton-by-George* By *M & H*	Means ...... Distance *5/-*	Text..	*17*	Dated
MM. Instructions.		Collected ...	Instns.	*7*	
M H 5/- paid. Northampton		Paid out *5* *0*	Total	*35*	
Handed in at the *Northampton* Office at *5 · 52 P*.M., Received here at *6 · 13 P*.M.					Stamp.

357. *The transmitted message.*—The clerk who receives an 'X' message writes it down in pencil on the proper form, which is indicated to him by the prefix. The prefix, code time, number of words, time received, are entered upon it, and the form is then passed on to the next circuit, upon which it is further signalled, and there it is dealt with as a forwarded message. The entries made on the B form are as follows :—

B. POST OFFICE TELEGRAPHS. No. of Message ....

		Words.		MM. Instructions.	Office Stamp.
Prefix *X*	Recd. at *6.0 P*M.	*11*	Sent at *6.13 P*M.	*M. H. 5/- Paid*	
	From *N H*	*17*	To *S. A.*		
Code *EKS*	By *J. Brown*	*7*	By *F. Smith*	*Northampton*	

358. *Acknowledgment.*—Upon the A B C and needle instruments every word is acknowledged as it is received. If a word cannot be read repetition is desired, in the former instrument by giving the letter R, in the latter by giving the letter E. If it be correctly read then it is acknowledged in the former instrument by a complete revolution to *zero*, in the latter instrument by the letter T. Expert operators are very fond of receiving without giving any acknowledgment until the end of the message.

In the Morse and Sounder, when worked by single currents, no acknowledgment is given until the end of the message, unless an evident error is made or the receiver fails to read, when the sender is instantly stopped and requested to repeat. When worked by double currents no acknowledgment is given until the message is complete ; when, as in single current working, if the total number of words received agrees with the total number signalled and the message appears accurate, the name of the addressee of the message, followed by the repetition of figures and doubtful words, is given as the acknowledgment. If the numbers do not agree, or if an error is evident, correction or repetition is obtained. When messages are very long it is sometimes the practice in double current working to take an acknowledgment occasionally to find that all is going well. On such occasions the switch is simply turned and the receiver sends 'G,' which means 'right, go on' if all is right, and 'R (repeat) from ——" if anything is wrong.

On very busy circuits worked by experienced operators, acknowledgments are frequently dispensed with altogether and the messages are pushed through as fast as clerks can manipulate. Of course, if a correction or repetition is needed the sending operator is stopped at once, but if no notice is given the sending clerk assumes that all is right, and he goes on manipulating until he is stopped by want of messages or by some necessity at the receiving end.[1]

When numbers are expressed in figures (§ 57) frequent errors are made in their transmission from the great similarity of the signals used. They are, therefore, repeated back, *collated* as it is called, by the receiving clerk as a part of the acknowledgment. But in automatic and duplex circuits the repetition is given by the forwarding clerk at the end of the message. Accuracy is thus ensured.

The acknowledgment on the continent is generally the repetition or *collation* of the numbers or cyphers signalled in the body of the message, and if there be no numbers the repetition of the number of words only.

359. *Modes of Working.*—The ordinary mode of working is to send messages consecutively in their order of precedence as indicated by their prefixes and code time, and stringent measures are adopted to prevent any abuse or neglect of code time; but when circuits become crowded with messages, special modes have to be devised to clear off the work. Thus, if we were to take a busy circuit at one extremity of

[1] On one occasion 613 messages were sent in one continuous stream upon an automatic circuit, between Goodwood and London, without one stop.

England, say from Plymouth to Penzance, if all the messages forwarded from and received at Penzance were to take their code turn, the forwarded messages would be considerably delayed by the received messages which arrived with earlier codes from other parts of England, and to which they would have to give way. This is remedied sometimes by adopting what is called *up* and *down* working; that is, by each station sending alternately one or several messages. When only one message is sent from each station alternately it is up and down working proper. When several messages are sent together in one direction it is *batch* working. Five is the favourite number for a batch. Of course each station adheres to code time in its own messages, and each station acknowledges each batch on its completion. When the business between two stations admits of it, the greatest amount of work is got off by having separate and distinct up and down circuits, along which the messages are constantly pouring in one continuous stream; and the perfection of this mode of working is obtained in the duplex circuits, where the stream is never interrupted by corrections, repetitions, or conversation. The duplex and automatic circuits are worked on the same method: messages are sent in batches and acknowledged in batches.

A proper system of working in batches is very desirable in cases of breaks-down, where, perhaps, one wire will have to do the work of three or four. Supposing there are four busy stations, A, B, C, and D, reduced by accident to using one wire, thus:—

Then to get the greatest work out of the wire with the least possible delay each station should in its turn send and receive a batch not exceeding five messages. Thus B sends its batch to A, C then sends its batch to A, and D follows; after this A sends its batch to B, then to C, and lastly to D, and so on. Of course, the circuit must be well controlled by the principal station A, and every clerk, who must be of the most experienced class, must be on the watch to take his turn without delay.

360. *Tablet check.*—It is always desirable to know what amount of work every circuit is doing, and the rate at which this work is being got off. For this purpose every circuit at most of the principal stations is supplied with a form called a *Tablet check*, upon which each message as it goes off is ticked—

Dated Stamp.

POST OFFICE TELEGRAPHS.

Bristol .... Circuit.

Hours	5 minutes and under	6 to 10	11 to 15	16 to 20	21 to 30	31 to 45	over 45	Total For^{d}.	Re-ceived	Total Rec^{d}.	Total	Remarks
9 to 10	////////	//	/					11	////// ////	10	21	
10—11	///////// ///////	///	/					20	////// //////	12	32	
11—12	////////// ///////	/						18	////// ////// ///////	19	37	
12—1												
1—2												
2—3												
3—4												
4—5												
5—6												
Total												

After each message is signalled, the forwarding clerk compares the code time, or time received in the case of a transmitted message, with the time sent, and he makes a tick or stroke upon the tablet form in the column for the delay, corresponding to the number of minutes of difference between these two times. Thus, if the message be received at 11.52 and sent at 11.54, the delay is *two minutes*, and the tick is made in the first column. If the message had not been sent until 11.59 the delay would have been *seven minutes*, and the tick would have been made in the second column, and so on for each column.

After each message is received a tick is made in the column for received messages, the delay upon these messages not being indicated. At the end of each hour all the ticks for the forwarded and received messages are added up and inserted in the proper columns, and at the end of the day all the vertical columns are added up and entered in a book or upon a form and filed away for future reference. A *constant* record of the amount of work done as well as of the delay on every circuit in an office is thus maintained with the least possible consumption of time on the part of those who keep the record.

361. These forms are summarised upon a return which is sent to the officer in charge of the District every eighth day, who, by taking an average of six or eight of these returns, is thereby acquainted with the amount of work transacted upon every circuit under his charge, and the manner in which this work is got off. He is able thereby to provide for the wants of the service, to see where the shoe pinches, to prevent congestion of the wires, and to supply the circuit accommodation needful for the District.

362. The essential secret of success in telegraphic working is constant personal supervision, strict and determined discipline, and extreme regularity and tidiness in offices. The greatest source of annoyance to those in charge of telegraphs is the loss of messages arising from careless distribution or collection, from messages falling behind instruments, getting improperly sorted, getting blown away by the wind in draughty offices, &c. If, however, the rigid rule of having a place for everything and everything in its place, especially upon the instrument counters and floors, be attended to, such accidents cannot arise. In a well-conducted office not a clerk is away from his instrument, not a trace of paper is on the floor, not a pad or a form is out of its place. The

distributors and collectors move about silently collecting and distributing their messages. The superintendent, or his assistant if the office is big enough to employ one, maintains every circuit under his constant supervision, relieving pressure here, checking circulation there, providing for a breakdown in another place, and fostering that swift transmission of messages which is the criterion of good telegraphy. In England our criterion of good working is this, that each station shall get off 90 per cent. of its work under ten minutes, and 50 per cent. under five minutes. Many stations actually get off 90 per cent. under five minutes. Thus, on August 30, 1875, Brighton, with 1,536 messages, got off 94 per cent. within this period, and most stations exceed the criterion. On August 19, 1875, Bristol, with 4,621 messages, got off 75 per cent. under five minutes, and 97 per cent. under ten minutes.

363. All messages as they are handed in are paid for in stamps or money; they are recorded in a book called the abstract book, and at the end of the day are collected together, with all those received and transmitted at the station, and sent up with a daily cash account to the General Post Office, where they are examined, checked, and filed away for future reference. All inland messages are kept for three clear months, after which period they are sent to the paper makers to be converted into pulp.

INDEX.

www.ingramcontent.com/pod-product-compliance
Lightning Source LLC
LaVergne TN
LVHW020235110826
845151LV00003B/930

* 9 7 8 1 4 2 5 5 3 0 1 5 0 *